# 52 MORE THINGS YOU SHOULD KNOW ABOUT PALAEONTOLOGY

EDITED BY ALEX CULLUM & ALLARD W MARTINIUS



Agile Libre

Agile Libre

First published in 2017 by Agile Libre
Nova Scotia, Canada. *www.agilelibre.com*



*Technical editors* Alex Cullum & Allard W Martinius
*Publisher* Matt Hall • *Managing editor* Kara Turner
*Indexer* Isabel Steurer • *Series design* electr0nika



The front cover incorporates an *Asteroceras obtusum* carving by Alex Cullum. This ammonite lived in the Jurassic ocean about 190 million years ago, and the specimen Alex based his carving on was collected at Lyme Regis, UK.

**Library and Archives Canada Cataloguing in Publication**

52 more things you should know about palaeontology / edited by
Alex Cullum & Allard W. Martinius.

Includes biographical references and index.
ISBN 978-0-9879594-7-8 (softcover)

1. Paleontology. I. Cullum, Alex, 1969-, editor II. Martinius, Allard W., 1963-, editor III. Title: Fifty-two more things you should know about palaeontology.

QE711.3.F52 2017 560 C2017-902300-4

# Who we are

Agile Libre is a small independent publisher of technical books in Nova Scotia, Canada. This book is part of a series; other books in the series include *52 Things You Should Know About Palaeontology*, *52 Things You Should Know About Geophysics*, *52 Things You Should Know About Geology*, and *52 Things You Should Know About Rock Physics*. Our passion is for sharing, so our books are openly licensed and inexpensive to buy. We hope they inspire you.

# Where to get this book

You will find this book for sale at *agilelibre.com*, and also at Amazon's various stores worldwide. Professors, chief geoscientists, managers, gift-givers: if you would like to buy more than 10 copies, please contact us for a discount at *hello@agilelibre.com*.

# About open licenses



# Colophon

This book was laid out on a Mac using Adobe InDesign. The cover typeface is Avant Garde Gothic and the text typefaces are Minion and Myriad. The figures were prepared in GIMP and Inkscape. It is printed and distributed through Amazon's CreateSpace publishing platform.

# Contents

## Alphabetical



### Essays

# Contents

## By theme

### Theme: FUNDAMENTALS

**Tags:** FUNDAMENTALS



### BIOLOGY

MORPHOLOGY • ECOLOGY • ETHOLOGY

## EVOLUTION

EVOLUTION • PHYLOGENY



## SPECIMENS

CURATION • COLLECTING

## METHODS

METHODS • LABORATORY • FIELDWORK • DESCRIPTION



## APPLICATION

APPLIED • BIOSTRATIGRAPHY

## PROCESS

TAPHONOMY • ICHNOLOGY



## CULTURE

CULTURE • STORIES • HISTORY • PHILOSOPHY

# Introduction

Following on from the first collection published in 2015, this collection of 52 more essays presents a wonderfully wide anthology of topics in palaeontology and explanations of individual species and taxa. Each of the contributions displays the unbridled enthusiasm of the author, with a desire to understand the biology and ecology of earlier life and its role in our biosphere. The essays also bear witness to the strength of the contributors' abilities to communicate scientific findings in a clear, enthusiastic, and unpretentious manner while not over-simplifying the science. If you are not an avid palaeontologist already, there's a danger you might become one while reading these essays. So, a big 'thank you' to all the contributors who shared with us their wonder with past natural life, and their adventures in discovering these insights. It has once again been a joy working with you.

This book takes us on a discovery journey through time, through animal kingdoms and palaeogeographies to reveal a wide range of characters. From the fantastic rediscovery of tiny Palaeocene frogs in the middle of Bombay, to Balearic mouse goats, snakes with four legs, and the beautiful insects preserved in Cretaceous amber. It illustrates how far our analytical techniques, based on indirect information, have developed by contrasting the legend of the Virgin Mary ascending a vertical rock face in Portugal with the exciting application of the rapidly developing field of extracting genomic information from extant organisms. In the Carboniferous, it takes us out on a swim in a shrimp-infested and thermally stratified freshwater lake, and a walk through arborescent swamp forests dominated by lycophytes. According to one of the contributors, you don't need to venture further than your mother-in-law's bathroom to study some of the world's finest palaeontology.

It is with a note of sadness that we dedicate this volume to three eminent palaeontologists who passed away over the last two years. Martin Brasier left a truly remarkable legacy most importantly related to the evolution and palaeoecology of the Ediacaran to early Cambrian organisms. David M Raup, an innovative and influential palaeontologist and expert on evolution and mass extinctions, introduced statistical concepts to palaeontology, and his work opened avenues in the studies of extinction patterns and biodiversity. Håvard Selnes was a truly outstanding applied palynologist whose offshore support to drilling teams is greatly missed. They left a legacy and our work continues to build on their achievements.

*Alex Cullum & Allard W Martinius*
*Norway, July 2017*

## A NOTE FROM THE PUBLISHER

We're excited to welcome a second palaeontology title to the *52 Things* family. The new collection Alex and Allard have compiled contains some modern gems which I'm certain will make you smile, and I hope will make you think.

After supporting the Educational Trust of The Micropalaeontological Society with the first *52 Things... Palaeontology*, we turn this time to supporting vertebrate researchers. We're donating $2 from the sale of each book to the Society of Vertebrate Paleontology. We hope this helps them in their mission to advance the science of vertebrate palaeontology around the world.

*Matt Hall*
*Nova Scotia, July 2017*

# A case of mistaken identity

**Tony Doré & Michael Oates**

From the dawn of civilization, humans have been picking up fossils and pondering what they could possibly mean. Up until the Age of Reason (mid-seventeenth to mid-eighteenth century) very few considered that fossils could ever have been alive. And why should they? There was no concept of geological time, and how could an animal have been turned to stone, except perhaps by an encounter with a Gorgon? The Danish anatomist Nicholas Steno was one of the first to recognize that fossils had once been living creatures, when in 1666 he connected curios known as 'tongue stones' with modern-day sharks' teeth.

The realization that fossils were ancient life forms brought a slew of misidentifications which, although hilarious today, were perfectly understandable at the time. The organic and inorganic were routinely confused, so innocent concretions became 'stone hearts' and 'flint feet', while belemnite guards were thought to be 'thunderbolts'. One of the most celebrated examples, actually part of the femur of the predatory dinosaur *Megalosaurus*, was first described by Robert Plot in his *Natural History of Oxford-shire* (1677). Plot speculated that the bone came from a race of giants that once populated the earth, but the physician Richard Brookes clearly had a very specific part of the anatomy in mind when, in 1766, he reclassified the specimen as *Scrotum humanum*. Lest we chortle too much, let's not forget that it would be another 60 years before the first dinosaur — fittingly, *Megalosaurus* — was described by William Buckland.

In those days, it was pretty easy to make mistakes about function too. Henry de la Beche's famous drawing *Ancient Dorsetshire* (1830) — rightfully celebrated as one of the first attempts to envisage a palaeoenvironment — shows ammonites sailing serenely on the surface like Spanish galleons amongst the general Jurassic carnage. The notion seems to have persisted in William Evans, Lord Energlyn's *Through the Crust of the Earth* (1973), which has the recumbent bivalve *Gryphaea* floating like a boat, while ammonites become 'ancient submarines'. A regular Jurassic flotilla…

It's also possible to perceive signs of life based on no evidence whatsoever. We've all done it at some time or other. Stare at a page of random dots for long enough, and eventually you start seeing complex shapes. This human need to discern patterns in meaningless information has even got a name: apophenia. Its greatest geological exponent was Randolph Kirkpatrick (1863–1950), an assistant keeper at the British Natural History Museum. Kirkpatrick became obsessed with the large foraminifera called nummulites, and started seeing their characteristic concentric patterns everywhere, including in igneous rocks. Full-blown monomania took over when, in 1912, he published *The Nummulosphere*, a monumental work that more or less postulates that the entire Earth is made of nummulites.

So does Kirkpatrick take away the coveted 'Master of Self-Delusion' prize? Not even close! Step forward, one Lewis A Manson. His masterwork, *The Birth of the Moon* (1978), published by a university press, displayed powers of misinterpretation so great as to leave us breathless with admiration. Whilst also reinventing most of astronomy and atomic physics, his refreshing take on geology included a new era, the Diastrophic. This era lasted but 96 hours, during which time a passing star dragged off enough crust to form the moon, opening faults into which all sea life disappeared, to be converted to oil 60 miles down. When these chasms slammed shut, huge quantities of sand spurted out, together with the petrified remains of molluscs' fleshy parts (bits of flint to you and me). Manson's lavish illustrations also identify a fibrous gypsum specimen as a piece of coconut 'with the fibres perfectly preserved' and a copper ore concretion as a petrified frog (the green colour is a dead giveaway).

So why are we fascinated by this stuff? Is it just because we like to feel superior? Not at all — these books would be among the first we'd rescue from a burning building. The ability to be spectacularly wrong, shrug your shoulders and move on, is one of the joys of working in an imprecise science. We might even admit to a few howlers of our own, although fortunately space considerations do not permit us to go into those right now.

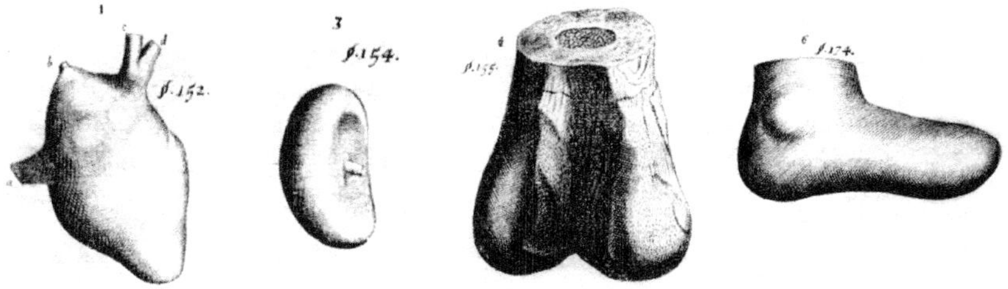

Some etchings from chapter 5 of *The Natural History of Oxford-shire* (Plot 1677) showing a heart, a kidney, a thigh bone — later described as a scrotum by Richard Brookes — and a foot.

## References

Energlyn, Lord (William David Evans) (1973). *Through the crust of the Earth*. McGraw-Hill. 127 p.

Kirkpatrick, R (1912). *The Nummulosphere: An Account of the Organic Origin of So-called Igneous Rocks and Abyssal Red Clays*. Lamley & Co. 424 p.

Manson, L A (1978). *The Birth of the Moon*. Dennis-Landman Publishers, 205 p.

Plot, R (1677). *The Natural History of Oxford-shire*. 358 p.

# A closer look at fossil sex

## Benni Bomfleur & Hans Kerp

Reproduction is a paramount feature of life. In order to maintain genetic diversity and promote evolutionary change, most plants and animals use sexual reproduction, and have been doing so throughout their evolutionary history. Indeed, among the many peculiar palaeontological discoveries of recent years are several reports of rather direct fossil evidence of sexual reproduction. The more explicit examples of this kind of '*in flagranti* preservation' include, for instance, several pairs of fossil turtles from the famous Eocene Messel Lake in Germany in which the male is attached in an unambiguous position behind the slightly larger female (Joyce et al. 2012). The ill-fated lovers must have plunged down into nocuous deeper waters while mating, eventually suffocating and sinking into the muddy lake bottom to become fossilized together. There are several other cases in which mating pairs of various types of insects have become forever caught in the act, engulfed by sticky tree resin or buried instantaneously by volcanic ash-falls (Li et al. 2013).

But there are other modes of fossilization that provide us with even more intimate details. One particularly bizarre kind of palaeontological treasure trove has only just been recognized. Clitellates — the group of worms that includes earthworms and leeches — secrete decay-resistant egg-cases that can become fossilized quite easily. Fossil clitellate cocoons are therefore not uncommon in certain sedimentary rocks of Mesozoic or Cenozoic age. The solid wall layers of these cocoons may in some cases contain fossilized microorganisms that became trapped and encased in the sticky cocoon material before it had fully hardened. But more remarkably, these cocoons have recently been shown to also preserve leftover bits and pieces of the animal's sperm (Bomfleur et al. 2015). Distinctive anatomical features of the sperm cells may even yield important clues to the identity of the cocoon producers, which may in turn help resolve the evolutionary history of worms and leeches, which have otherwise left hardly any trace in the fossil record.

But surprisingly, the most graphic sexual imagery of the fossil record comes not from animals but from tiny plants. The famous Early Devonian Rhynie Chert Lagerstätte from Scotland is a hot-spring deposit that contains an early terrestrial ecosystem from a time when plants were still rather primitive and

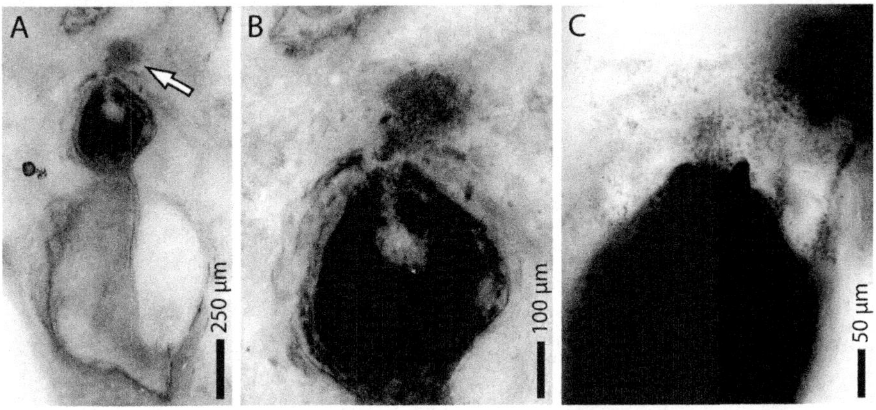


Intimate moments captured in chert: the tiny clouds (arrow) are fossilized sperm ejected from the male sex organs of early land plants preserved in the Devonian Rhynie chert, Scotland. To the left is *Remyophyton delicatulum* (A, B), the gametophyte life form of the *Rhynia gwynne-vaughanii* plant; to the right (C) is *Lyonophyton rhyniensis*, the gametophyte life form of *Aglaophyton major*.

growing only up to some decimetres tall. Rapid silicification from hydrothermal fluids preserved entire stands of this miniature vegetation three-dimensionally and in phenomenal subcellular detail. Among other wonderful oddities like fossilized lichens, mycorrhizae, hatching nematode babies and one of the world's oldest fossil penises (Dunlop et al. 2003), the Rhynie Chert has also recorded the complete series of life-history stages of some of the earliest land plants, including the delicate and rarely preserved, sexually reproducing gametophyte generations (Kerp et al. 2003). The images above show the male sex organs of two common Rhynie Chert plants in which contact with silica-rich fluids caused the antheridia to pop open and eject their sperm into the solidifying fossilization medium.

Not only do such fossil finds attract considerable media attention (sex sells after all even after millions of years), they give us crucial insights into the evolution of reproductive behaviour and may have an important bearing on systematic placement and relationships. What's more, the glamour of such fossil curiosities carries a special lesson of its own: the fossil record never ceases to surprise.

## References

Bomfleur, B, T Mörs, M Ferraguti, M Reguero and S McLoughlin (2015). Fossilized spermatozoa preserved in a 50-Myr-old annelid cocoon from Antarctica. *Biology Letters* **11**, 20150431.

Dunlop, J, L Anderson, H Kerp, and H Hass (2003). Preserved organs of Devonian harvestmen. *Nature* **425**, 916.

Joyce, W, N Micklich, S Schaal, and T Scheyer (2012). Caught in the act: the first record of copulating fossil vertebrates. *Biology Letters* **8**, 846–848.

Kerp H, N Trewin, and H Hass (2003). New gametophytes from the Early Devonian Rhynie Chert. *Transactions of the Royal Society of Edinburgh: Earth Sciences* **94**, 411–428.

Li, S, C Shih, C Wang, H Pang, and D Ren (2013). Forever Love: The Hitherto Earliest Record of Copulating Insects from the Middle Jurassic of China. *PLoS ONE* **8**, e78188.

# All in a day's work

**Mike Whitaker**

The lounge on this oil rig is basic but boasts an endless vista of grey sea. Just now it's filled with bright orange survival suits and bodies trying to squeeze themselves into toughened body bags. We are the selected few to fly off on the next helicopter.

I am a palynologist and part of the biostrat team here on the rig, leaving behind two dedicated technicians, Vigdis and Pia. Work on board is much like in any other laboratory. We work our shifts in a small lab housed outside in a steel container. They take small samples, usually every 30 m, from the nearby, noisy, and fume-filled shaker room, in which the cuttings carved by the drill bit are recovered from the drilling mud. Preparation includes a 30-minute immersion in hydrochloric and then hydrofluoric acid before the resulting organic residue is mounted on a glass slide. I can then observe the assemblage of dinocysts, spores, and woody debris through my microscope.

I had been on the rig for three weeks, it was time to go. During that time most of my waking hours had been spent staring down a microscope looking for clues as to the exact layer of the Brent Group the drill bit was supposed to be drilling. Indeed, was it even within the Brent Group? Unfortunately I had spotted

Håvard Selnes working at his microscope in an A60 cabin on the Heidrun field.

extreme forms of *Nannoceratopsis gracilis angularis* which I knew marked the uppermost layer of the underlying Drake Formation. We were drilling below the Brent Group and had not encountered reservoir sands or hydrocarbons. This was not good news. A re-think was needed by the geology team. Despite this disappointment, my job was done and it was time to hand over to Håvard Selnes who would arrive on the incoming helicopter. He is a top-class palynologist, excellent laboratory technician, and skilled communicator. I was pleased to know the unit would be in his capable hands.

The long-awaited call '10 minutes!' came and we heard the sound of the helicopter landing. Everyone made their way, bags in hand, to the helideck. The strength of the wind could easily blow a person off the deck and into the sea, but somehow we held on, dumped our bags, and climbed aboard the helicopter. The doors closed as the motors roared ready for take-off. It's always a nervous moment: a fierce gust of wind could down the helicopter into the sea or crash it against the rig. But the take-off went well. Since the well was located north of the Snorre Field at about 61.5°N, it would be a long flight back to Bergen.

Then I noticed something strange. The windows in the helicopter seemed to be getting increasingly obscured with dirt. As I watched, the windows became virtually covered in what I now realized was oil. The interior of the helicopter became distinctly warmer and I was beginning to sweat. The smell of diesel became intense. Engine noise prevented any audible communication as confusion grew. In the cockpit the pilots seemed unalarmed. Meanwhile, those next to me were unzipping their survival suits and others were actually taking off the top part of the suits as the temperatures climbed.

It was clear the helicopter was on fire and the windows were now completely black with oil. But it continued to fly without any visible sign of being in trouble. The pilots never moved from their seats nor did they turn around to see the growing chaos behind. We continued in this state of nervous fear for about 10 minutes before temperatures began to cool. Nothing further happened. As we landed, the pilots dryly announced that the engine had caught fire but they had switched successfully to the back-up engine. As we left the helicopter and entered the terminal building nothing else was said. All was calm as though nothing had happened. It was all in a day's work.

# An inordinate fondness for beetles

**David Peris**

*There is a story, possibly apocryphal, of the distinguished British biologist, J B S Haldane, who found himself in the company of a group of theologians. On being asked what one could conclude as to the nature of the Creator from a study of his creation, Haldane is said to have answered, "An inordinate fondness for beetles."*

**G Evelyn Hutchinson (1959)**

Haldane's wry observation reflects the high diversity of species within the insect order Coleoptera which, with almost 400 000 described species, exceeds that of any other known group of animals or plants. In fact, almost one in every four known animal or plant species on Earth is a beetle.

Why there are currently so many beetles, and why this group of insects has had such evolutionary success in the past, are questions many researchers have tried (and are trying) to answer. Still without consensus, it was these mysteries which prompted my PhD project, completed in 2015. I had a key source of evidence: the possibility to look through 'a window to the Cretaceous': amber.

Amber is the result of the fossilization of plant resins, which have the property of preserving animal or plant inclusions with a high level of detail. There are several hundred outcrops providing amber worldwide (Martínez-Delclòs et al. 2004; Grimaldi & Engel 2005, and George Poinar Jr in this volume); although only some of them preserve bioinclusions. Amber has resulted in an exceptional resource in palaeontology because it conserves not only the external morphology of the fossils, but also their behaviour in many cases, which is scarcely preserved in rock contents (Peris et al. 2017). As the insect trapped within represents a frozen moment in time, it is common to find reproductive behaviours (such as mating, egg lying, or progeny care), social behaviours (such as swarming or food carriage), predation, parasitism, mutualism, and defensive behaviours (Arillo 2007). Thus, the fossil record of beetles in amber provides both an understanding of their evolution and an explanation of the modern assemblages we can observe today.

While Baltic (Eocene) amber and Dominican and Mexican (Miocene) ambers are internationally known because of their commerce, Cretaceous amber is much rarer and highly coveted by researchers. Beyond being a time when dinosaurs were not yet extinct, it is the period when flowering plants appeared and radiated, displacing the great forests of gymnosperms which had dominated the land until then. The Cretaceous was therefore a time of change, making the study of Cretaceous beetles important for a number of reasons:

1. Coleoptera are one of the better represented orders of insects in amber deposits worldwide, and those from the Cretaceous have only been intensively studied in recent years — making them an under-examined group.
2. Their study will provide information about ancient ecosystems, but also increase our knowledge of the recent forms through the phylogeny of the basal groups. There appears to be a turning point in Mid-Cretaceous fauna: the amber record from the Early Cretaceous (for example from Lebanon and Spain) show fossils defined as a new supra-generic taxon, with characters intermediate between two (or more) recent groups which makes their classification very difficult. Immediately following, in the amber record from Mid to Late Cretaceous (for example those from France and Myanmar) there are fossil beetles which can be assigned unequivocally to recent groups (Peris et al. 2016).
3. Beetles have been identified as having played a major role in the pollination of flowering plants. With the absence of major current pollinator insects — such as bees or butterflies — until the Late Cretaceous, beetles and flies have generally been considered as the hypothetical first pollinators of flowering plants during the Early Cretaceous (Friis et al. 2011). Current studies are focusing on the link between beetles and the success of the angiosperms versus gymnosperms through the Early Cretaceous (Peris et al. 2017).

Beetles trapped in amber surely provide evolutionary biologists with the ideal system to discover the exciting story of their evolutionary success.

## References and acknowledgments

Arillo, A (2007). Paleoethology: fossilized behaviours in amber. *Geologica Acta* **5**, 159–166.

Friis, E, P Crane, and K Pedersen (2011). *Early flowers and angiosperm evolution*. Cambridge University Press. 585 p.

Grimaldi, D and M Engel (2005). *Evolution of the insects*. Cambridge University Press. 755 p.

Hutchinson, G (1959). Homage to Santa Rosalia, or why are there so many kinds of animals? *American Naturalist* **93**.

Martínez-Delclòs, X, D Briggs, and E Peñalver (2004). Taphonomy of insects in carbonates and amber. *Palaeogeography, Palaeoclimatology, Palaeoecology* **203**, 19–64.

Peris, D, E Ruzzier, V Perrichot, and X Delclòs (2016). Evolutionary and paleobiological implications of Coleoptera (Insecta) from Tethyan-influenced Cretaceous ambers. *Geoscience Frontiers* **7**, 695–706.

Peris, D, R Pérez-de la Fuente, E Peñalver, X Delclòs, E Barrón, C Labandeira (2017). False blister beetles and the expansion of gymnosperm-insect pollination modes before angiosperm dominance. *Current Biology* **27**, 897–904.

This essay is part of the AMBARES (CGL2011-23948) and AMBERIA (CGL2014-52163) projects, funded by the Spanish Ministry of Economy and Competitiveness.

# Antarctica's *Glossopteris* forests

**Stephen McLoughlin**

The Prince Charles Mountains (PCMs) of East Antarctica were first spotted from the air during The United States Navy Antarctic Developments Program 'Operation Highjump' in 1946–47, but first visited nearly a decade later by Australian explorers. In the austral summer of 1994–95, Andrew Drinnan (University of Melbourne) and I ventured to the PCMs to search for fossil plants. Most of the rocks there are ancient — Archaean and Proterozoic granites and metamorphics — but one small area, the Amery Oasis, hosts Permian coal-rich strata and Triassic red-beds.

The Permian strata include one especially interesting coal seam of which the top 40 cm constitutes a permineralized peat. This layer of peat was impregnated by silica so rapidly that the original organic matter was entombed in three-dimensions before it could be compressed and coalified. When thin-sectioned for microscopic examination, the peat layer reveals exquisite anatomical details of the plants that grew in the 260 million-year-old swamp. Not only are the robust woody tissues of the plants preserved, but also delicate leaf cells, together with the hyphae and spores of fungi and water moulds that were actively decomposing the forest litter at the time of burial. There are even signs of animal life in the form of fragments of insect exoskeletons and the traces of arthropod activities. Some pieces of fossil wood contain regular holes around the growth rings that are filled with coprolites — the fossil dung left behind by the seasonal feeding activities of voracious mites. Other coprolites contain the pollen and spores of ferns and seed plants. Yet others contain the spores of fungi or fragments of leaf tissue denoting a diverse array of specialized feeding habits. The peat also contains bands of charcoal left by forest fires that scorched the Permian swamp vegetation. The peat layer is overlain by laminated siltstones indicating that the ancient swamp was eventually drowned by deepening lake waters. All these signs of physical and biotic interactions are now enabling us to piece together the architecture of the ancient coal swamp organisms and the complexity of their high-latitude ecosystems (Slater et al. 2015).

Leaves of *Glossopteris*, an extinct seed-plant characteristic of the Permian (300–252 Ma), are by far the dominant fossils in this deposit. The curious distribution of *Glossopteris* leaves across the Southern Hemisphere continents and

India was central to early arguments that these regions were once united in a vast supercontinent — Gondwana (McLoughlin 2011). Glossopterids were substantial trees, perhaps reaching 30 m tall, with roots containing distinctive chambers that provided an aeration system for growth in waterlogged soils. The *Glossopteris* plant was the main contributor of organic matter to the formation of coal, not only in the PCMs, but across the entire Southern Hemisphere in the Permian. The economies of India, Australia, and South Africa now heavily depend on extraction and use of these vast coal reserves.

Typical *Glossopteris* leaf in Permian lake beds overlying the permineralized peat in the PCMs.

In the PCMs, as in other parts of Gondwana, *Glossopteris* fossils persist until the uppermost Permian strata — then abruptly vanish. Thus, *Glossopteris* was one of the major casualties of the end-Permian biotic crisis, an event that Doug Erwin of the Smithsonian Institution has described as the mother of all mass extinctions. The ultimate decline of *Glossopteris* may have spanned only a few thousand years, before being replaced by new groups of small- and thick-leafed hardy conifers and advanced seedferns. The immediate causes for the abrupt demise of the vast *Glossopteris* forests are written in the overlying Triassic strata. The transition from coals, deposited under consistently moist conditions in the Permian, to red-beds, characterized by haematite staining, dessication cracks, and calcrete nodules indicative of semi-arid climates in the Triassic, indicates dramatic drying and warming across the temperate regions of Gondwana. Conditions simply became too dry and hot for *Glossopteris* to persist. Strikingly, the signs of warming and drying are evident throughout the globe at that time (McLoughlin et al. 1997). What constituted the trigger for such runaway heating and drying is a topic for another article, but it is pertinent in this time of modern global warming to recognize just how rapidly dominant plant groups and productive continental ecosystems can disappear under a changing climate.

## References

McLoughlin, S (2011). *Glossopteris* – insights into the architecture and relationships of an iconic Permian Gondwanan plant. *Journal of the Botanical Society of Bengal* **65**, 93–106.

McLoughlin, S, S Lindström, and A Drinnan (1997). Gondwanan floristic and sedimentological trends during the Permian-Triassic transition: new evidence from the Amery Group, northern Prince Charles Mountains, East Antarctica. *Antarctic Science* **9**, 281–298.

Slater, B, S McLoughlin, and J Hilton (2015). A high-latitude Gondwanan lagerstätte: The Permian permineralised peat biota of the Prince Charles Mountains, Antarctica. *Gondwana Research* **27**, 1446–1473.

# A snake with four legs

**David Martill**

Everyone knows snakes don't have legs, and that other reptiles do.

Arthropods were first to acquire their legs in the Cambrian. They first exercised them out of water in the Silurian. Vertebrate limbs made their first appearance in the Devonian, and evolved in fully aquatic animals. By the Carboniferous, four-limbed vertebrates were recognisable as amphibians and reptiles; by the Permian there was a group well on the way to becoming mammals. Four legs were proving extremely useful on land, and their form and function has evolved greatly through geological time.

A few animal groups have lost just one pair of limbs. Loss of the rear limbs is common for secondarily aquatic animals, manatees doing this in the Late Eocene, while cetaceans (whales and dolphins) hung on to their rear limbs until the Miocene. It seems ironic that limbs developed in aquatic animals, but were later lost in response to a return to the watery world.

Losing one pair of limbs is unfortunate, but to lose both pairs seems downright irresponsible. Surprisingly though, the list of limbless tetrapods is quite long. Most famously this list includes the snakes, but added to it are the ceacelian amphibians, and several legless reptiles including almost all amphisbaenians and several lizard species. The giant skinks of South Africa and the slow worm of Great Britain are lizards that cope remarkably well without any legs. So it is not lack of legs that makes a snake, well, a snake. Snakes are identified by a whole suite of characters that make them snakes, and loss of limbs isn't one of them.

Snakes have lots of unique features, including a single row of belly scales and they shed their skins in their entirety. They have a transparent, immovable cover over their eyes, so they don't need to blink. Snakes eat other vertebrates whole, and have unfused mandibular symphyses and backward-pointing teeth which allow them to ratchet very large prey items down their gullet. Their bodies are longer than their tails, and their vertebrae have additional articulation surfaces that enable them to both curve their bodies but also hold them very stiff. Many aspects of their skull anatomy suggest that they are most closely related to monitor lizards, and may have evolved from them or shared a common ancestor with them, probably splitting off in the Jurassic. If this scenario

*Losing one pair of limbs is unfortunate,*
*but to lose both pairs seems downright irresponsible.*



is correct, then there should be some ancestral snakes with legs. Indeed, fossil two-legged snakes have been found in Cretaceous strata. These two-legged snakes possess rear limbs and have been found in the Mediterranean region and South America, including *Eupodophis* from the Lebanon, *Pachyrachis* and *Haasiophis* from Palestine, and *Najash* from the Late Cretaceous of Patagonia.

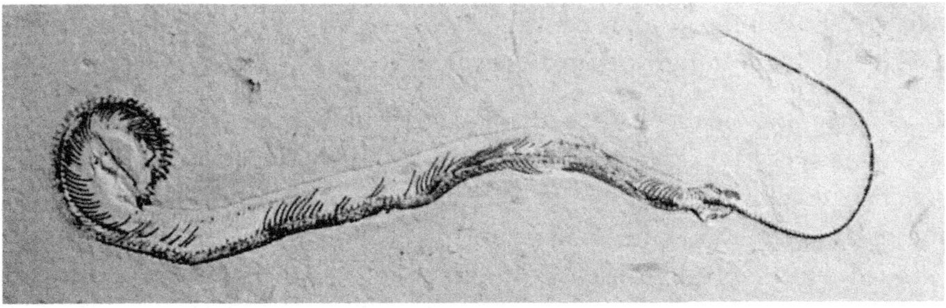

*Tetrapodophis* from the Cretaceous Crato Formation of Brazil (total body length 195 mm).

These remarkable fossils are clear evidence for the early possession of limbs in snakes, but an even more exciting discovery was made recently when a four-legged snake was discovered. *Tetrapodophis amplectus*, from the Early Cretaceous Crato Formation of Brazil is a small fossil that looks very much like a snake. Close inspection reveals not only small hind legs, but also a pair of extremely small forelimbs. This curious little fossil with four legs would not be classified as a snake except for the fact that it also has impressions of the single row of belly scales, it has a tail that is shorter than its body, it has backward-pointing teeth, and it has vertebrae with extra articulatory surfaces. Just to clinch the deal, the stomach contents of this little fossil are preserved, and within the mass of material are tiny bones showing that it was a carnivore. While this unique little fossil is not considered to be the ancestor of all modern snakes, it is the smoking-gun evidence that snakes once possessed four legs. *Tetrapodophis* is on display at the Bürgermeister-Müller-Museum in Solnhofen, Germany.

# A sticky end

## George Poinar Jr

There are three main liquid exudates from plants: resins, saps, and gums. Saps are chiefly water-soluble fluids containing sugars and minerals. Resins and gums are viscous secretions that can harden, but the latter are water-soluble while resins are water insoluble. The preservative qualities of plant resins have been known for centuries — humans have used resins as embalming agents, for treating wounds, and for preventing wine spoilage. Animals and plants in fossilized resin, commonly known as amber, represent some of the most exquisitely preserved remains of extinct creatures dating back millions of years.

During maturation, many tree resins, including those from both gymnosperms and angiosperms, harden into amber and function as excellent preservatives that maintain the original structure of the inclusions. The preservative qualities are, in large part, due to the immediate action of the resin on the organs, tissues, and cells of entombed invertebrates, vertebrates, and plant remains.

Certain qualities of plant resins make them excellent preservatives. The methodology involved in the preservation of amber inclusions involves four major actions:

- **Infiltration.** As the resin infiltrates over and throughout the entrapped inclusions, it restricts oxygen and releases antimicrobial compounds that destroy decay-causing microorganisms that could decompose the tissues. The degree of infiltration depends on the size of the inclusion, if it is completely immersed, the thickness of its outer covering, and any openings either natural or produced as a result of a trauma.
- **Fixation.** Preserving the organs, tissues, and cells of entrapped victims, fixation is accomplished by natural products in the resin, such as various terpenoids and resin acids.
- **Dehydration.** This is accomplished by sugars and terpines in the resin that replace water within the victim's tissues, resulting in a condition similar to the type of physical preservation known as inert dehydration. In brief, the dehydration processes produce an extreme case of mummification, very similar to a natural embalming process.
- **Polymerization.** After fixation and dehydration, the slow process of polymerization and cross-bonding of the molecules begins as the resin hardens.

*By infiltrating the tissues and cells of the entrapped organisms, [plant resins] effectively freeze time for millions of years.*



This process also protects the inclusions from environmental factors that occur during the deposition and often re-deposition of the product. During this polymerization process, a semi-fossilized material known as copal is produced after several thousand years. Amber is eventually formed when the state of fossilization of the resin reaches a hardness of between 2–3 on the Moh's scale, a refractive index between 1.5 and 1.6, a specific gravity between 1.06 and 1.10, and a melting point between 250–300°C. The process of maturation of resin into amber takes millions of years. Most authenticated amber with identifiable inclusions dates from 15 to 130 million years.

The preservative qualities of plant resins are astonishing. By infiltrating the tissues and cells of the entrapped organisms, they effectively freeze time for millions of years.

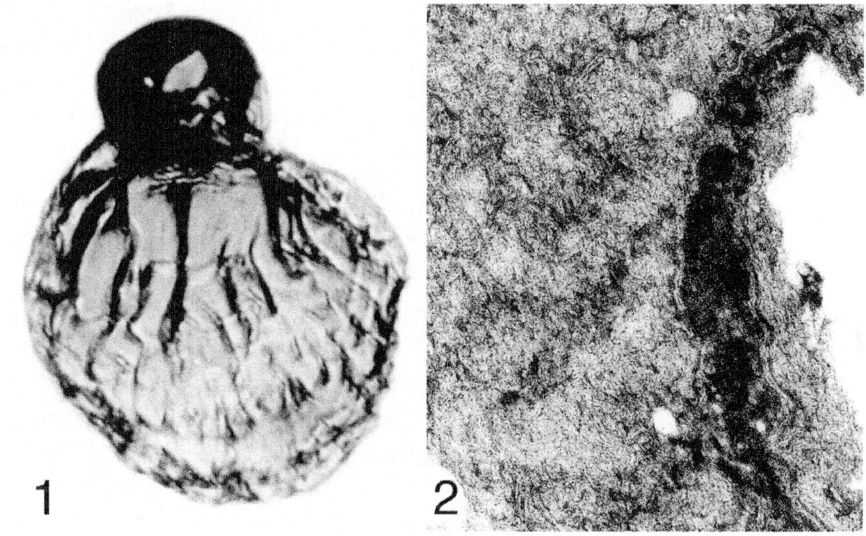


1. The 100 million-year-old protozoan *Microrhopalodites polynucleatis* in the mid-gut of the termite *Kalotermes burmensis* in Myanmar amber. Length of specimen is 164 microns. 2. An ultrastructural image of a nucleus (note dark nucleoplasm surrounded by a nuclear envelope) in hypodermal tissue inside the abdomen of a fungus gnat (Diptera: Mycetophilidae) in 40–50 million-year-old Baltic amber. The length of the nucleus is 3.5 microns. Photo by Roberta Poinar.

# A taste for bones

**Steffen Kiel**

During a deep-sea expedition in 2002, a strange worm that appeared to be living and feeding on whale bones was discovered at a depth of nearly 3000 m off the coast of California. It was named *Osedax*, the bone-eater. Soon, further species were found in the Atlantic Ocean and around Japan, and were presented as 'wonderful worldwide whale worms' at a conference at the Scripps Institution of Oceanography, which was when I first heard about them. Being a palaeontologist interested in the evolution of deep-sea ecosystems, I wondered whether *Osedax* had been around before the whales, perhaps feeding on the bones of Mesozoic marine reptiles, and most importantly, how to find fossil evidence for an organism that has no hard parts such as a shell or skeleton?

Trace fossils left by the worm's feeding habit emerged as the way forward. *Osedax* bores chemically into bone, leaving a tiny hole on its surface, and then extends a spongy 'root system' into its interior to feed on its collagen matrix, dissolving the bone in the process. I sent images of some bored bones to a friend who had spent decades collecting whale fossils in western Washington State, USA. This seemed just the right place: plenty of uplifted deep-water sediments of roughly the same age as the first whales.

Indeed, only two days later my friend told me that a small packet was in the mail, containing the rib fragment of a fossil whale with suspicious holes. This fossil would later be called 'the shabbiest fossil that ever made it into a dignified scientific journal' by an Austrian press agency. Faced with the problem of documenting tiny trace fossils hidden in a brittle bone fragment, I turned to a colleague who ran the micro-computed tomography lab at my institute. Micro-CT uses hundreds of x-ray images at a microscopic scale to produce 3D images of the object. And what we saw on these images looked exactly like individual specimens of living *Osedax*.

Of course we were not alone in our research. A British–Italian team was hunting for the same trace fossils, and while our manuscript was in review, a Spanish team published alleged *Osedax* borings from shallow marine sediments in southern Spain. But these borings were not quite right — they were too long, too straight, and lacked the characteristic spongy root system. Plus, they were geologically too young to be of evolutionary significance. In contrast, our

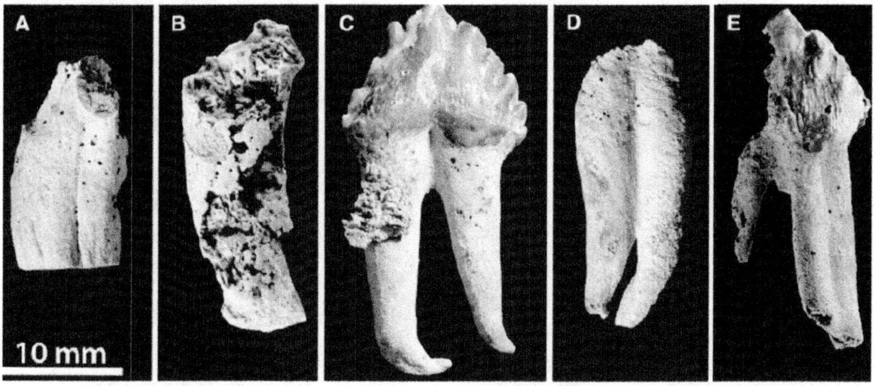


Teeth of two mysticete whales with *Osedax* borings, Oligocene, Washington State, USA. A and B are from the latest Oligocene part of the Lincoln Creek Formation. Specimens C–E are from the Late–Early Oligocene part of the Psyht Formation. Note that the *Osedax* borings are restricted to the upper (presumably exposed) part of the teeth in C and D. From Kiel et al. 2011.



rib fragment with the very *Osedax*-like traces was about 30 million years old. This indicated that *Osedax* had been feeding already on the earliest open-marine whales.

But what about those large marine reptiles of the Mesozoic? There were two problems here: first, until then *Osedax* was only known from whale bones. Second, even if *Osedax* had lived on Mesozoic reptiles, what had it been feeding on after their extinction and before the origin of whales, a timespan of at least 25 million years? Here we were ahead of the biologists. In the same deep-water sediments in Washington, we discovered traces of *Osedax* in the bones of marine diving birds, showing for the first time that *Osedax* was able to feed on non-mammalian bones. Marine birds were around in the Mesozoic and had survived the end-Cretaceous mass-extinction event, and could thus have provided a continuous food source for *Osedax*.

Since then, there has been fossil and biological evidence for *Osedax* feeding even on fish bones, so the idea of the 'whale-bone specialist' was put to rest. And eventually, the British–Italian team found *Osedax* traces in Cretaceous marine reptiles. These fossils — a plesiosaur and a sea turtle — stored for more than a century in the palaeontological collection of the University of Cambridge, provided the ultimate evidence that *Osedax* existed during the Age of the Dinosaurs, when it was happily feeding on reptile bones.

## Acknowledgments

Kiel, S, W-A Kahl, and J Goedert (2011). *Osedax* borings in fossil marine bird bones. *Naturwissenschaften* **98**: 51–55.

# Birds of a fibula

**Jon Tennant**

Over the last 20 years, there has grown insurmountable evidence that birds are the direct modern descendants of dinosaurs. Eagles are dinosaurs. Pigeons are dinosaurs, annoyingly. Even penguins are weird, swimming dinosaurs.

The supporting data comes from a whole range of scientific domains, from the discovery of thousands of feathered dinosaurs in the fossil record to chemical and biological analysis of these fossils at a molecular level. As science progresses in terms of technical capability, what we can glean from the fossil record increases too, and we learn more about the link between dinosaurs and birds.

A rapidly advancing research field at the moment is in embryonics and genetics. You might ask, What does this have to do with dinosaurs? Well, it's a good point. We don't have any dinosaur DNA, and we have very few fossilized embryos. But wait! If modern birds are dinosaurs, then that means we actually have a lot of dinosaur DNA and their embryos! This means we can start asking a whole range of new questions about the evolution and origins of birds from dinosaurs.

Recently, an international team of researchers took on the task of looking at developmental links between dinosaurs and modern birds (Botelho et al. 2016). There is an old and quite controversial saying that 'ontogeny recapitulates phylogeny', which means that the growth or development of an individual can be a reflection of the evolution that has occurred along its lineage. Wouldn't it be cool if we could see that in birds and their dinosaur ancestors?

By looking at the development of the fibula (one of the lower leg bones) of chicken embryos, the team discovered that the far end of the fibula expresses a gene known as Indian hedgehog (great name!) during development. This gene plays an important role in ossification — the replacement of cartilage with bone.

The Indian hedgehog homolog (IHH) gene works as part of a feedback system with another gene, not-so-nicely named parathyroid hormone-related protein (PTHrP), which acts to delay bone growth and also inhibits the further production of IHH. So if PTHrP is activated, it leads to decreased growth of the bone.

The development of this feedback system in the chicken embryo seems to be related to a reduction in the size of the fibula, compared to an earlier embryonic stage in which both the fibula and tibia start out at similar lengths to each other. This

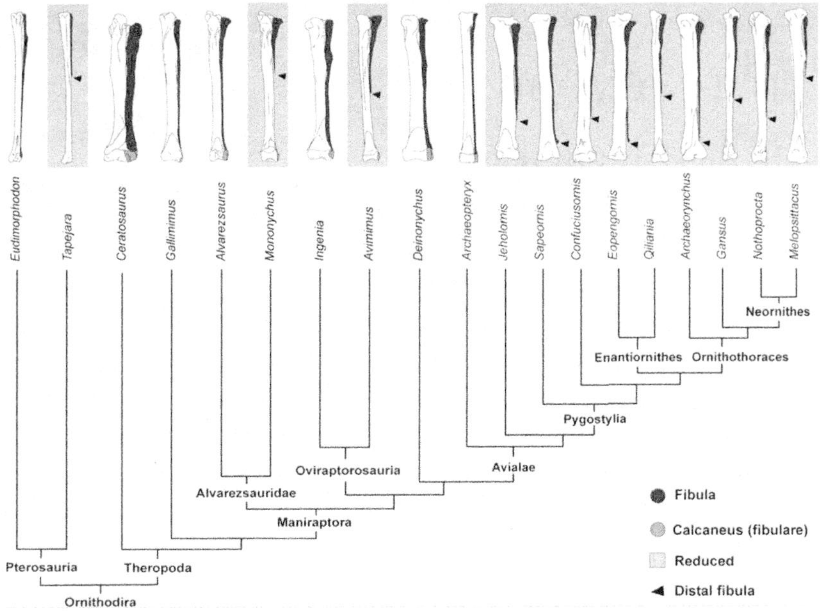


The convergent reduction of the distal fibula in Ornithodira, the clade that contains pterosaurs and dinosaurs.

is really neat because if you look at fully grown birds, they have a much smaller fibula, shaped like a splinter, compared to the tibia (the other lower leg bone).

Almost all birds from the Mesozoic era also have a fibula that is much shorter than the tibia, and does not even touch the ankle bones any more. In some early birds, including the now extinct Enantiornithes, the fibula was still almost as long as the tibia. The story the fossil record tells us is one of the formation of an early splinter-shaped fibula, followed by the independent and convergent acquisition of the reduction of this bone in several distantly related early bird groups, including the one that led to modern birds.

This means that dinosaurs and early birds shared the same or similar pattern of fibula development, reflected in their evolutionary relationships and through time. The reason why this happens remains a bit of a mystery. Modern birds of different sizes and ecologies all show evidence of this fibula reduction. This suggests that it is what is called a 'non-adaptive' process, as it is highly unlikely that such a feature would play a part in such different roles.

We're only just beginning to unlock the molecular links between dinosaurs and birds, and this represents a fascinating glimpse into the future of this research field. Stay tuned!

## References

Botelho, J, D Smith-Paredes, S Soto-Acuña, J O'Connor, V Palma, and A Vargas (2016). Molecular development of fibular reduction in birds and its evolution from dinosaurs. *Evolution*. March 2016. 543–554.

# Convergent and parallel evolution

## Enrico Savazzi

Virtually every course on palaeontology and evolution includes a section on convergent and parallel evolution. The two phenomena result in two or more taxa displaying similar sets of characters. To qualify as convergent or parallel evolution, morphological similarity alone is not sufficient. It must be accompanied by similar function. Classical examples are the wings in birds, bats, and pterosaurs, and the body shape in fast-swimming sharks, porpoises, and ichthyosaurs.

According to the classical definition, parallel evolution takes place among closely related taxa (e.g. two species within the same genus). Convergence, instead, takes place among distantly related groups with a long history of separate evolution (e.g. two taxa in different classes). If the genetic affinities are intermediate between these extremes, assigning examples of morphological similarity to either parallel or convergent evolution entails a degree of subjectivity. It is therefore preferable to use convergence as an umbrella term for all evolution leading to morphologically and functionally similar results, and restrict parallel evolution to convergence among closely related taxa.

Evolution does not take place in a genetic or functional vacuum. All organisms are related, since they share the same biochemical basis for storing, replicating, and expressing genetic information. Selective pressures, in similar conditions, can be expected to lead to similar adaptations. Indeed, studies in constructional morphology (Savazzi 2014) show that convergent evolution is the norm, rather than a remarkable oddity as envisaged by earlier palaeontologists and biologists.

In principle, it is possible for a retrovirus to transplant working genes from one species to a distantly related one, through random excision of genetic information from the source and random insertion into the recipient. This, however, may be a rare event in nature. A comparable process is used in genetic engineering, but requires the carrier virus to be carefully engineered in order to increase the odds of success to acceptable levels. Aside from this, there is no possibility for a species to 'copy' the evolutionary achievements of an unrelated one.

Thus, virtually all instances of convergence among distant taxa are likely the result of selective pressure independently acting on the genome of each of these

*There is no copyright on evolutionary innovations in nature, and no patent office...*

taxa, i.e. independent evolution (although morphogenetic building blocks may be shared even among distantly related taxa). There is no copyright on evolutionary innovations in nature, and no patent office to prevent a species from evolving a set of functional characters similar to those already developed by another species. When an emergent species directly competes with an already established one, exploiting the same adaptations and ecological niche, this may be an obstacle to success of the former species. A different geographic location, environment or geological age prevents such competition. This is seen, for example, in the convergent evolution of antifreeze proteins in Arctic versus Antarctic fishes (Chen et al. 1997).

Parallel evolution within a taxon can be facilitated by a common set of genes shared by multiple species, including species where these genes are present but not phenotypically expressed. This makes parallel evolution likely to be common, albeit difficult to prove. The definition of parallel evolution implies that, somewhere along the ancestry lineage that joins two phenotypically convergent species, at least one intermediate species lacks the character in question. If the whole lineage consistently displays the character, instead, this is not parallel evolution but just plain inheritance.

## References

Chen L, A DeVries, and C-H Cheng (1997). Convergent evolution of antifreeze glycoproteins in Antarctic notothenioid fish and Arctic cod. *Proceedings of the National Academy of Science of the USA*, **94** (8): 3817–3822.

Savazzi, E (2014). Constructional morphology and morphodynamics. In: Cullum, A and Martinius, A (eds.): *52 things you should know about palaeontology*: 38–39. Agile Libre, Nova Scotia.

# Corals meet reefs in the Carboniferous

**Markus Aretz**

I study corals and reefs around the world. You are probably thinking what most people do at this point, but no it's not just jetting from one tropical paradise to another, drinking cocktails as the sun sets. Please let me explain.

I am in the northwest of Ireland. It's cold and rainy, and the waves of the Atlantic Ocean are rolling against the cliff I am studying. It's early November, but there is no real difference with my last field campaign here in early June. So why aren't I at a nice resort in the tropics?

The answer is simple. I'm studying splendid outcrops on the southern side of Donegal Bay exposing thousands of fossil corals and an ancient ecosystem from Early Carboniferous times when a shallow tropical sea covered this part of the world. For a palaeontologist and sedimentologist studying Carboniferous corals and environments, this area *is* paradise.

However, there is a slight set-back. There are no true coral reefs here. Now, before you tell me to pack my bags and head to the tropics, we have to think a little bit about the history of fossil reefs and the organisms that built them.

Reefs are almost as old as the life on our planet. But we have to wait until the beginning of the Phanerozoic before metazoans contributed to reef building; before that it was only microbial communities. Corals form today's reefs, but this doesn't imply that fossil corals did the same thing during the entire Phanerozoic. Yes, they did their fair share of the work, but the geological record shows that reef building was quite popular among a number of organisms during different periods. Some are rather exotic such as the rudist bivalves which dominated in the Cretaceous.

The reef systems of stromatoporid sponges and corals collapsed during two mass extinctions in the Late Devonian. With the Devonian reef builders out of the picture, the Carboniferous saw a number of organisms in different combinations try to re-establish the art of reef building. Corals are among them, but appear to be quite slow learners.

Creating a wave-resistant structure like a reef requires reef-builders that are firmly attached to the substrate or else encrust it. Corals possessing this ability

Streedagh Point section in northwest Ireland showing a patch of the large solitary corals *Siphonophyllia samsonensis* within a coral meadow. Note strong changes in the growth direction of some specimens; most likely after having been toppled during storm events.

got wiped out in the Devonian extinctions, and Carboniferous corals were at best only loosely attached to the substrate. Hence, ecological limitations prevented these corals from building true coral reefs in high-energy settings.

So if the corals alone weren't capable of building reefs in the Carboniferous, how did the Carboniferous reefs form that are, for example, found in Queensland, Australia? The answer is microbes again. They are not only the glue for sticking everything down, but also form massive bodies at the heart of the reef, acting as corals do in modern reefs. Many metazoans successfully contributed to Carboniferous microbe-metazoan reefs; corals are among them, as are sponges and bryozoans. But a Carboniferous reef without microbes just didn't work.

Modern coral reefs are very special ecosystems and they need our attention. However, when looking into the geological record, we have to open our eyes and not name everything a coral reef because there was once a coral living in a reef. Look for the framework builder and you will discover a completely new and exciting world.

# Decapods and domestic geology

**Alun Williams**

It's not always necessary to travel to exotic places to see nice rocks. One of the best limestone 'outcrops' I have seen in the UK is located in my mother in-law's bathroom. The walls of this room are covered in beautifully honed slabs of a creamy-white rock, which on closer inspection turns out to be a cemented grainstone (calcarenite), composed of near spherical coated-grains which may or may not be ooids (without a hand lens it's hard to say). They also contain a generous supply of small bioclasts that include mollusc shells, solenoporaceans, calcareous sponges, and solitary corals. But for me the most captivating features within the tiles are a series of dark patches, roughly elliptical in shape, that I have seen described in similar tiles as 'fossils' or even 'corals'. But they are neither of these, instead they are something more exciting: burrows, that are filled with a distinctively coarse-grained sediment bound by calcite cement.

Carbonate sedimentology is my specialist subject, and so when I finally put my mind to it I didn't have too much difficulty in tracking down the likely origin of this rock. I concluded that the rock is most likely from Portugal — more specifically from the Maciço Calcário Estremenho, north of Lisbon — where pristine limestones of Middle Jurassic age provide some of the world's most popular ornamental stone. Descriptions of the composition of these rocks (from publications and wholesalers) fit well with the bathroom outcrop, and they are similar to Middle Jurassic rocks from other parts of the world with which I am more familiar.

The age of the rocks is significant, because it was not until the Mesozoic era that decapods — crabs, shrimps, and lobsters — became the dominant burrowers in shallow-marine tropical environments. Modern-day sediments deposited in such regions are riddled with the homes of these creatures, ranging from simple Y-shaped burrows to complex gallery systems of interconnecting tunnels. They are so abundant that they commonly obliterate primary sedimentary structures (such as ripples), and form the dominant control on the texture of the sediment. These burrows can significantly modify the properties of the rocks that control fluid flow — an important consideration to petroleum geologists and hydrogeologists. (My colleague Dirk Knaust as well as Kevin Cunningham of the US Geological Survey have written some particularly good publications on this topic.)

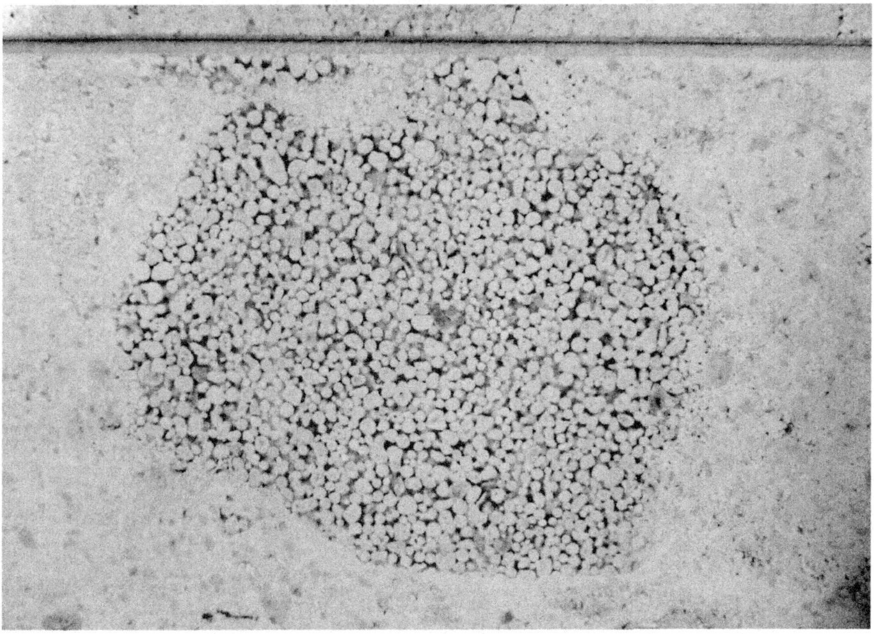

A close-up of a burrow, showing the fill of well-rounded coarse-grained sediment with calcite cement between grains, as found on a bathroom tile.

The tiles in my mother-in-law's bathroom are interesting in this regard, as they consist of rocks deposited in high-energy, sand-dominated settings. And these are not easy places to make burrows, as anyone who has tried to dig a hole on a sandy beach can testify — above the water table it's not too hard, but below it the hole keeps filling with sediment. So the burrow-makers, which are most commonly shrimps, need to do a lot of work to keep their burrows open. They plaster the walls with a sticky layer (can you guess where they get this substance from?), they use rocks, corals or large shells to shore up the roof, and they have to constantly remove sand grains that fall into the burrow. It's fun to watch the shrimps on modern-day sea floors, where many of them live in symbiotic relationships with little gobies (fish) who act as their 'bodyguards' (this is in itself a fascinating area of study). And the fossil traces of these creatures are preserved in countless bathrooms, staircases, and hotel lobbies around the world, unnoticed by millions of people who pass by them every day.

## References

Cunningham, K, M Sukop, Huang, Haibo, P Alvarez, H Curran, and R Renken (2009). Prominence of ichnologically influenced macroporosity in the karst Biscayne aquifer: Stratiform "super-K" zones: *Geological Society of America Bulletin* **121** (1–2), 164–180.

Knaust, D (2014). Classification of Bioturbation-related Reservoir Quality in the Khuff Formation (Middle East): Towards a Genetic Approach. In: Poppelreiter (ed.), *Permo-Triassic Sequence of the Arabian Plate*, EAGE, 247–268.

# Deep in the mire at Yons Nab

## Peter H Morris

The imposing Middle Jurassic cliffs of the North Yorkshire coast have been subject to palaeontological studies for over 100 years, so you could be forgiven for thinking that little was to be gained from further research. But then most of the sections are non-marine and previous studies were mainly palaeobotanical, focusing on the plant-bearing horizons, and despite the recent boom in the North Sea's Middle Jurassic (Brent Group) oil and gas production, only limited comparative micropalaeontological analyses had been undertaken. Because of this, and driven by my desire to apply the discipline more rigorously, one December morning I set about logging and sampling the Gristhorpe Member outcrop at Yons Nab, near Scarborough.

Analysis of the basal unit established the marine–brackish trend from the occurrence of agglutinated foraminifera. Above this level, determining the change-over to non-marine deposition had always been problematic since much of the plant debris and associated palynomorphs were considered to have been transported. It was here that new light was about to be cast: a succession of carbonaceous claystones yielded increasing numbers of megaspores culminating in a half-metre-thick, super-abundant horizon, packed with diverse species. Jurassic megaspores can be up to 1 mm in diameter and were produced by small herbaceous lycopsid plants having two types of sporangia, housing megaspores and microspores that were initially dispersed in fresh water. Species present included *Erlansonisporites sparassis*, *Horstisporites kendalliae*, and *Paxillitriletes phyllicus*, with the total specimen count exceeding 4000 per 100 g of sediment. Was this the 'smoking-gun' of a climax lycopsid flora that had colonized a lake-mire habitat? I thought so, because associated with it were numerous megaspore tetrads, abundant miospore masses, and high

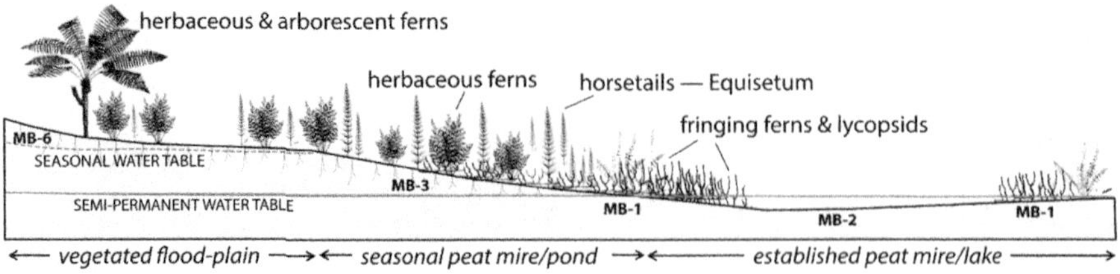


A reconstruction of the profile across the Gristhorpe Member delta plain showing the plant habitats associated with the lake-mire at Yons Nab. MB represent micro-biofacies types, with MB-1 and MB-2 being megaspore-rich.

*Driven by my desire to apply the discipline more rigorously, one December morning I set about logging and sampling the Gristhorpe Member outcrop at Yons Nab.*

levels of plant debris, all indicating this flora lived close by. More evidence was needed however, so David Batten then analyzed the palynofloral content of the samples collected.

On comparing megaspore and miospore distributions, a high positive correlation with filicopsid spores was evident, notably those of *Klukisporites*, derived mainly from mire-fringing ferns. This backed up the megaspore evidence for development of a lake-mire habitat. I asked myself many questions about this lake-mire, such as how much time and space had it occupied, and in what way did its floral composition vary? The Gristhorpe Member was deposited over 100 000 years so it could be deduced that its claystones accumulated at an average rate of 19 years/mm giving a duration of about 11 000 years for the lake-mire bed. The bed can be traced for 300 m in outcrop at Yons Nab, with the possibility that it correlates with the 'Coal Marker' 14 km north at Cloughton Wyke. There, however, the character of the mire and associated claystones was different, with a marked reduction in megaspores and *Klukisporites*. To explain this difference we proposed that the water-table there fluctuated seasonally, resulting in a decrease in megaspores/spores from lycopsids and fringing-fern floras (see figure) with a corresponding increase in background gymnosperm pollen.

Recognition of the lake-mire deposits at Yons Nab was a breakthrough in our understanding of the controls on megaspore and small-spore generation and dispersal. It was pivotal in developing a biofacies-based model which could be used to accurately determine brackish–freshwater trends and changes in groundwater levels through Middle Jurassic fluvio-deltaic cycles (see Morris & Batten 2016). So be like the Saxon-gold detectorists in Staffordshire and apply new tools to old locations — undiscovered treasures await!

## References and acknowledgments

This essay was written with contributions from David J Batten.

Morris, P and D Batten (2016). Megaspores and associated palynofloras of Middle Jurassic fluvio-deltaic sequences in North Yorkshire and the northern North Sea: a biofacies-based approach to palaeoenvironmental anlysis and modelling. *Journal of Micropalaeontology* **35** (2), 151–172.

# Dinosaur evolution: it's about time

## Denver Fowler

We are living in a golden age of stratigraphy. Advances in radiometric dating let us date rocks with unprecedented precision. We use sequence stratigraphy to track depositional accommodation cycles basin to basin, correlating across continents like never before. Stratigraphy has come of age, and it is revolutionizing how we can study dinosaur evolution. I know what you're thinking: 'That's great, but what does this mean for dinosaurs?'

Consider if you were studying the evolution of North American Cretaceous dinosaurs, and were trying to radiometrically date volcanic ash layers that occurred in the same formations as your fossil bones. A 1980s researcher would have had a choice of potassium–argon (K–Ar) or uranium–lead (U–Pb) methods, and could retrieve the age with an error of maybe ± 1 or 2 million years. Good enough to get a rough idea of which units might be similar in age, but really not much better than the resolution available using mammal biostratigraphy (five million years or so).

Things have got much better since the late 1980s. The new technique of argon–argon (Ar–Ar) dating, and refinement of U–Pb methods, mean that both can retrieve radiometric dates with errors of just one or two hundred thousand years, sometimes even less. However, there have been some problems: U–Pb dates typically came out ~1% older than Ar–Ar. Also, Ar–Ar analysis requires comparison to a mineral standard of known age, and different laboratories historically used different standards, with differing reliability; even the ages of these standards have been changing every few years as they are refined!

Unfortunately, palaeontologists (myself included) had been citing published dates without recalibrating or making sure the different standards, methods, or analyzed minerals were cross-compared correctly. This meant that many Late Cretaceous formations were miscorrelated by as much as a million years.

Thankfully, these days most geochronologists use the Fish Canyon sanidine standard, dated at 28.201 Ma, which also happens to reconcile the difference between Ar–Ar and U–Pb dates. OK, so now we can accurately correlate those formations. If we also have high-resolution stratigraphic data for dinosaur specimens (which is actually much rarer than you would think or hope) then what can this tell us about their evolution?

The author cleans mudstone from around a *Triceratops horridus* skull. Hell Creek Formation, Montana, 2016.

One interesting observation is that dinosaur taxa often occur as a succession of short-duration species which do not overlap stratigraphically. This is the expected pattern produced by lineages evolving through time without branching into multiple contemporaneous species, and suggests that what used to be perceived as diversity is probably better explained as rapid evolution in a much reduced number of lineages.

This has had knock-on effects. For example, one recent hypothesis proposed that dinosaurs might have had very restricted geographic ranges, limited to individual depositional basins. Revision of radiometric dates is showing that this might instead be an artifact of miscorrelation, showing species as contemporaneous when they were actually slightly separated in time.

Other advanced stratigraphic methods also tell us about dinosaur palaeobiology. In terrestrial sequence stratigraphy, accommodation cycles used to correlate geological units also encode subtle environmental differences occurring at the time of original deposition (such as variation in heights of water tables, prevalence of hydromorphic soils, and swamps versus more well-drained settings). Through careful analysis of stratigraphically controlled fossil data, we can observe fluctuations in relative species abundance concomitant with these cycles, reflecting individual species' habitat preferences for swamps versus not-swamps.

Finally, the finding that many dinosaur lineages evolve recognisably new morphologies in as little as 300 000 years or less means that dinosaurs might be the highest resolution biostratigraphic index fossils available for Late Cretaceous terrestrial units: even better than mammals. It's an exciting time to be a dinosaur stratigrapher.

# Dreaming of plant fossils

**Mike Pole**

Step one in researching plant fossils in New Zealand is often trying to find out who owns what, where they live, and then catching them at home. In rural areas a good first step is often the local pub. I had to do this just yesterday. It was mid-morning when I wandered in and the bar was deserted, so I kept heading in the direction of voices in the back. All the activity was in the kitchen. I found the boss and gave her my line about what I did.

The manageress walked back to the bar, pulled out a chair, leaned over its back and looked at me thoughtfully. 'So, when you were 10, did you lie in bed and dream about plant fossils?'

I told her she was very close to the mark. I had in fact, found my first leaf fossil at almost ten.

It is Easter 1971. The family is holed up in a tiny home-made caravan at Cole Creek, in New Zealand's South Westland just north of Haast. It's bucketing down. We're here because my Dad is obsessed with finding the missing backpack of a murdered woman, Jennifer Beard. The pack might just solve the case. But it is, and remains, a remote, rugged, sparsely populated area covered with forests and swamps. The backpack was never found and the case remains unsolved.

As the family descended into cabin fever, I popped out during a lull in the deluge to walk along the road cutting on the other (south) side of the Cole Creek bridge. I looked down at the rocks at my feet and at first glance thought what I was seeing was an exceptionally wet leaf, so soggy it had almost melted on to a rock. I then realized what it was — a fossil leaf on the bedding plane of a piece of shale. My first fossil leaf. The rain eased up enough for the rest of the family to emerge and discover a large shale boulder closer to the creek. As we split the boulder into layer after layer, a variety of other plant fossils emerged. There were leaves that looked like beech leaves and shoots that looked a bit like modern rimu and matai confiers. I, not yet 10, assumed they must be Carboniferous. This was, after all, the age of the great coal swamps elsewhere in the world.

More than 40 years later, the fossils remain unpublished, and the place where we chipped away at the boulder is now a young forest of New Zealand's beech tree, *Nothofagus*. I've pushed my way in a few times, but never found any more

My first leaf fossil: perhaps *Nothofagus*.

sign of fossils. The road cutting itself has weathered, and though more leaf fossils must remain, deep inside, there would be little chance that a 10 year old poking around today would find anything.

The rocks are usually mapped as Late Cretaceous, although I wonder if they may be Palaeocene. They lack obvious shoots of the Araucariaceae family (including New Zealand's kauri) which were common in the Late Cretaceous. There's an outside chance that the first leaf I found is related to *Nothofagus*, the tree that now covers the site, and is known to have been in New Zealand at the time. But most of those Cole Creek leaf fossils are unlikely to be any extant genus.

The manageress seemed satisfied. She pushed the chair back and said: 'Well, it gives you a goal. Something to aim for when you get out of bed.'

Indeed. And, to this day, something to dream about when I get in.

# Earth's own asteroids

**Liam Herringshaw**

Dinner is served. On a plate before you: a delicious roast chicken. The bird, however, is larger than your head and you have no hands or teeth, let alone cutlery. How are you going to eat it?

How about pushing one half of your stomach out through your mouth, dissolving the chicken in digestive juices, then hauling your belly back into place, full of nutritious bird broth? No? Well then, you're obviously not a starfish.

Asteroids, to give starfish their scientific name, are echinoderms, the group of marine animals that also includes sea urchins, brittle stars, sea lilies, and sea cucumbers. The five-fingered starfish is among the most familiar of all sea creatures; yet even a cursory investigation of their biology, ecology, and evolutionary history reveals this familiarity to be a deception. These icons of the intertidal zone are about as strange as life on Earth gets.

Echinoderms are pentaradial, having five-fold body symmetry. Their skeleton is made of uniquely porous calcite, with each little 'bone' — called an ossicle — growing as a single crystal. If attacked, many species can cast off a limb as a defensive measure and regrow it later. Some are even able to develop a whole new starfish from a single, sacrificed ray. The thought of losing your arm in an industrial accident only to see that arm grow into a twin brother is almost too bizarre to contemplate.

As thoroughly strange as all this might sound, it unequivocally works. The echinoderms are a submarine success story. The fossil record shows that asteroids have been doing their thing for about half a billion years.

Composed of tiny ossicles held together by soft tissues, a dead starfish has a fairly low preservation potential, but they do get fossilized occasionally. In the mid-19th century, a beautiful, slender-rayed specimen was found by a Mr J Richardson in the limestone beds of Point Riche, Newfoundland. Elkanah Billings named it *Stenaster huxleyi*, or Huxley's narrow-star.

Billings described the starfish as Silurian, rocks of that age being the oldest then thought to yield fossils. The starfish is actually from the middle Ordovician, during which time western Newfoundland was a tropical, shallow-water sea, with

'Huxley's original little star' — a fossil starfish discovered by J Richardson in the limestone beds of Point Riche, Port-aux-Choix, in the mid 1800s. Image from Charles Schuchert's *Text-book of Geology Part II: Historical Geology.*

reefs fringing the ancient North American continent of Laurentia. Unlike the present day, reef-building in the Cambrian and Early Ordovician wasn't accomplished by corals, but by even simpler creatures, mostly microbes and sponges.

In 1915, Huxley's narrow-star was renamed. American palaeontologist Charles Schuchert decided Billings had got his genus wrong, and so the asteroid became *Urasterella huxleyi* , or 'Huxley's original little star'. Schuchert also described the beds it had been found in as being the Chazy Reefs. These are fossiliferous Ordovician limestones that crop out in much of eastern North America, and they represent the first major flourish of animal-built reefs in the geological record.

What role were asteroids playing in these early carbonate mounds? Many living species are predators, but skeletal analysis of Ordovician starfish suggests they hadn't yet acquired the body and mouth flexibility that enables them to capture and devour mobile prey. It is more likely they were scavengers, or predators upon creatures that couldn't escape.

Ordovician starfish are scarce. Only one other specimen of *Urasterella huxleyi* has ever been found. For other clues of Ordovician asteroid activity, we must depend on trace fossils such as starfish resting impressions.

Whilst Schuchert's rebranding has stuck, it turns out that he was wrong about the age of Huxley's original little star. More recent research shows that the Point Riche reefs are slightly older than the Chazy ones, so they were truly pioneers of a brave new world.

# Foraminiferal microhabitats

**Frans Jorissen**

Foraminifera are traditionally studied by micropalaeontologists. After Joseph Cushman turned them into highly effective biostratigraphical tools in the 1930s, fossil foraminifera were also used to reconstruct palaeoenvironments. To better understand their palaeoecology, scientists started to observe foraminifera in modern ecosystems, explaining why even today studies on foraminiferal ecology and biology are commonly performed by geologists.

For a long time, it has been thought that foraminifera live at the sediment–water interface, where they would profit from the combination of oxygen-saturated bottom waters and abundant supplies of labile, easily consumable food particles. Only in 1985, in a seminal paper in *Nature*, Bruce Corliss showed that rose-bengal-stained, and therefore supposedly living, foraminifera were also present in deeper sediment layers, at least down to 15 cm. In fact, a succession of species was observed within the sediment. The concept of foraminiferal microhabitats was born.

In a follow-up paper, Corliss and Chen (1988) showed a clear relation between foraminiferal microhabitats and test (i.e. shell) morphology, making it possible to infer microhabitats for fossil species as well.

While I was finishing my PhD thesis on the ecology of recent Adriatic Sea benthic foraminifera, I read these papers with increasing excitement. A whole new world was pictured, with many yet undiscovered consequences. In fact, at the same time, a rather speculative paper appeared by Robert Carney (1989) on the biological functioning of deep-sea sedimentary environments. It hypothesized the existence of two completely decoupled cycles of organic matter degradation: a short, aerobic sediment surface cycle based on labile organic matter, and a much slower, anaerobic cycle within the sediment, based on much more refractory organic carbon.

Starting from this idea, Bert van der Zwaan, Henko de Stigter, and I wondered how we could apply this concept to refine the long-standing method of using the ratio between planktonic and benthic foraminifera in a sample to estimate palaeobathymetry. This works because the sea-floor faunas become less dense as the water depth increases. Our idea was that the ratio between planktonic forams and only the surface-water benthic forams should represent the ratio between productivity in the water and on the sea-floor better than the ratio between planktonic and all benthic foraminifera. With some examples, it was shown that this was indeed the case (Van der Zwaan et al. 1990).

Later research showed that foraminiferal microhabitats are essentially controlled by food supplies and oxygen concentration. The conceptual TROX-model shows how foraminiferal microhabitats may be explained by minimal oxygen penetration in eutrophic areas, but are probably due to an absence of food particles within the sediment in more oligotrophic settings (Jorissen et al. 1995).

The discovery of foraminiferal microhabitats has opened a whole range of new research pathways. For example, intensive research efforts have been made to understand how foraminifera could live in anoxic sediments, leading to the discovery of the facultative anaerobic metabolism of some taxa, capable of switching to nitrate respiration (Risgaard-Petersen et al. 2006). The particular resistance to low oxygen of these taxa has also led to the development of palaeo-oxygenation proxies, based on the relative proportions of these taxa.

Finally, the discovery of foraminiferal microhabitats has also revolutionized stable isotope geochemistry. By measuring the $\delta^{13}C$ (and trace-metal ratios) of different species, having calcified at different depths within the sediment, former redox conditions can be reconstructed, and a whole range of new proxies for palaeo-oxygenation and export production continue to be developed.

All thanks to the observation of a couple of rose-bengal-stained foraminifera in a sediment core!



## References

Carney, R (1989). Examining relationships between organic carbon flux and deep-sea deposit feeding. In: Lopez, G, G Taghon, J Levinton, J (Eds.): Ecology of Marine Deposit feeders. *Lecture Notes on Coastal and estuarine Studies* **31**, 24–58. Springer, Berlin.

Corliss, B (1985). Microhabitats of benthic foraminifera within deep-sea sediments. *Nature* **314**, 435–438.

Corliss, B and C Chen (1988). Morphotype patterns of Norwegian Sea deep-sea benthic foraminifera and ecological implications. *Geology* **16**, 716–719.

Jorissen, F, H de Stigter, and J Widmark (1995). A conceptual model explaining benthic foraminiferal microhabitats: *Marine Micropaleontology* **26**, 3–15.

Risgaard-Petersen, N, A Langezaal, S Ingvardsen, M Schmid, M Jetten, H Op Den Camp, J Derksen, E Piña-Ochoa, S Eriksson, L Nielsen, N Revsbech, T Cedhagen, and G Van Der Zwaan (2006). Evidence for complete denitrification in a benthic foraminifer. *Nature* **443**, 93–96.

Van der Zwaan, G, F Jorissen, and H de Stigter (1990). The depth dependency of planktonic/benthic foraminiferal ratios: Constraints and applications. *Marine Geology* **95**, 1–16.

# Fossil fishing needs patience

## Alison Murray

What do striding across the Western Desert of Egypt, climbing a mountain in Gansu Province of China, or wending a path through the rice paddies in Sumatra have in common?

The answer is that they are all top places for fossil fishing. Palaeontologists often get to visit far-off places to collaborate with local colleagues while searching for fossils. Images of palaeontologists baking in the desert sun while brushing sand off bones are prevalent on the internet and in movies, and we do in fact spend a fair amount of time doing just that in order to collect the often spectacular fossils we study. But there is also the less glamorous part of the job back in the lab.

Isolated fish bones may need little preparation in the lab; however, articulated fishes may require more time to remove the sediments and reveal the bones beneath. Like many palaeontologists with small material, and smaller budgets, I normally prepare the material myself. This has the added benefit of allowing me to see the details of the fish as I slowly work to remove the matrix, and gives me time to think about what animal it is and to which group it belongs. But sometimes, even after preparation, a fossil fish may remain an enigma. What to do then?

One such enigma was a partial, articulated fish recovered in part and counterpart from a freshwater Eocene site in Tanzania. The small specimen preserved the head and most of the body, but was lacking a tail. I made a latex peel of the two parts, and studied both the peels and the original specimens under the microscope. What could it be? There seemed to be no defining characteristics, and nothing to point me in the right direction. The specimen sat next to my microscope for over a year. Every few weeks, I would examine it again, and puzzle over what it might be.

My 'aha' moment when it came was not a moment at all, but more of a long slow painstaking trip to eliminate all the things it couldn't be. Due to the age and location, it was likely that this specimen would belong to a lineage of fishes that still inhabit Africa today. So I found a list of African fishes (*www.fishbase.org*), and worked my way through every African fish group. Some groups were easy to dismiss — eels and bichir are long and skinny, the predatory snakehead and bony tongue fishes have large teeth in the jaws, and the herrings have distinctive belly scutes. Eventually, I determined that the only thing this fish could be was a basal lineage of characiform (the group which includes the South American piranha and neon tetra). This process took many months, the main problem being that the fossil was missing the distinctive characiform caudal skeleton (a fusion between a couple of the bones supporting the fin rays and the compound terminal element of the vertebral column).



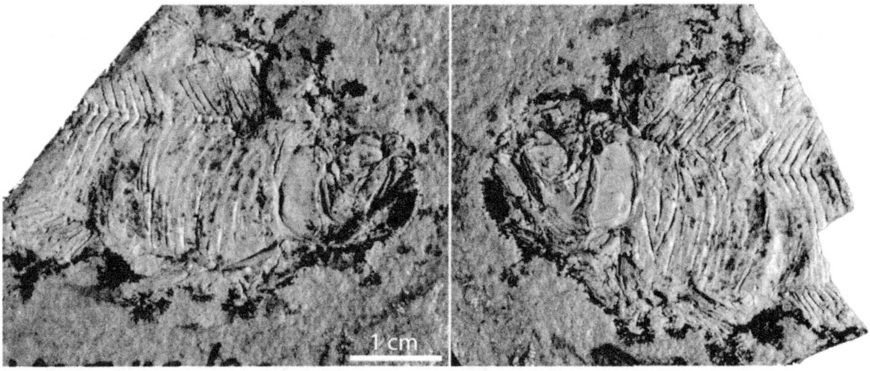


*Eocitharinus macrognathus* specimen and peel of the specimen under polarized light. The original specimen was returned to Tanzania.

However, when I tried to publish this fossil as a new species of characiform, one of the reviewers was not happy. I was told that I shouldn't just hastily name something without proper study and justification — my months of research were dismissed by someone who assumed I hadn't spent enough time on the project.

To respond to this charge, I worked my way through every fish group known, to show why my fossil could be nothing other than a characiform. I explained why it couldn't be a pike (there were no teeth on the roof of the mouth), or a perch (which have strong spines in the dorsal and anal fins), or even a salmon (they have two toothed bones forming the margin of the upper jaw, the maxilla and premaxilla; the maxilla is excluded from the gape in the fossil). In the end, after adding pages of justification for my identification, the little fish was published and named *Eocitharinus macrognathus*.

# Fossils for sale

## Tony Doré

Back in the late 1970s, there was an uproar in Britain's normally reserved palaeontological community. Commercial fossil hunters had arrived in the country with power tools and explosives and were excavating well-known fossil sites, including type localities and Sites of Special Scientific Interest (SSSIs). What bugged the palaeontologists was not just the apparent destruction of the sites, but also the fact that these people were out-and-out professionals, with no interest in the fossils save for their money-making capability.

A by-product of this invasion was that the localities now had spoil-heaps, full of fossil fragments discarded in favour of the perfect specimens. These reports came to the attention of me and my old college friend Mick Oates, so we decided to visit a few, and see what we could scavenge.

Our first visit was to the Silurian shales of Lesmahagow, near Glasgow. The locality is famous for well-preserved eurypterids. Sure enough the evidence of commercial activity was there, together with signs warning away future miscreants. More remarkable, however, was that the local authorities had excavated an order of magnitude more rock. Large quantities of fossiliferous shales had been packaged up into gabions (wire baskets filled with stones and used to reinforce roadsides) and the best *Slimonia* tail section found that day had to be carefully teased out through a wire mesh. Not quite what we'd been led to expect.

Next stop was Orkney, where the bituminous flagstones of the Devonian Sandwick Fish Bed contain beautiful lacustrine fossils, including some of the earliest lungfish. The main locality was a disused quarry, and the wicked exploiters had created new exposures where previously there had just been a weathered rock face.

A theme is emerging, is it not? Nobody wants to see an SSSI obliterated, but in order to have a classic fossil locality something has to create the exposure in the first place, whether or not we approve of the motives. In fact, it can be shown that many of the most critical fossils in the history of palaeontology came to us, not through the diligence of palaeontologists, but from opportunists looking for some extra income. The gentleman geologists of the 19th century, who made some of the biggest strides in the history of palaeontology, relied upon

Mary Anning of Lyme Regis, Dorset, and Mr. Wood's Fossils, Edinburgh.

workers in the quarries, building sites, railway cuttings, and canals who were only too eager to supplement their meagre incomes with a bit of fossil selling. The most famous 19th century fossil supplier was undoubtedly Mary Anning of Lyme Regis, Dorset, who supported herself and her family by selling fossils. She made some of the most important finds in palaeontology, including the first *Ichthyosaurus* and *Plesiosaurus*. Her patrons were among the biggest palaeontological names of the day such as William Buckland and Henry De la Beche.

Significant and sometimes groundbreaking contributions by the professional sellers have continued to this day. The title of 'modern-day Mary Anning' unquestionably belongs to Stan Wood, a self-taught fossil dealer and enthusiast from Edinburgh, Scotland. Unlike Anning, Stan could use a mechanical digger in the pursuit of the finest fossils. He eventually became a palaeontological celebrity through his many discoveries, which included one of the earliest reptiles, the Carboniferous *Westlothiana lizziae*. Stan passed away in 2012, but his business, Mr. Wood's Fossils, continues to thrive in Edinburgh.

So it seems there has always been a somewhat uneasy symbiosis between palaeontologists and the fossil sellers. But perhaps now, at the time of writing, the pendulum has swung too far in the commercial direction. At a reasonable price, any home can have one of those beautiful brown sectioned Madagascar ammonites (mostly *Cleoniceras*,) or a slab of those ubiquitous Moroccan nautiloids, polished beyond all recognition.

For me though, nothing comes close to finding a well-preserved fossil *in situ*. No mantelpiece specimen you can buy can replicate that sense of wonder and mystery.

# Giant lycophytes of the Carboniferous

**Barry Thomas**

*Stigmaria* are common plant fossils in Carboniferous age rocks. They are the large dichotomising rhizophore bases of Carboniferous arborescent lycophytes that formed the main plant cover of the huge tropical Coal Measures swamps. There was much early debate on the nature of the stigmarian axes until Brongniart gave the first detailed description in 1828 showing they were covered with spirally distributed roots or circular scars where the roots had once been.

The *Williamson Stigmaria* in Clayton quarry before its removal to Manchester.

The internal structure of *Stigmaria* was studied by a number of authors, but it was Williamson that published the first comprehensive anatomical study in 1887. Williamson purchased a complete *Stigmaria* from a quarry at Clayton, near Bradford, and moved it to Manchester where it is now on display in the Manchester Museum. Two other specimens from the same quarry were later moved to public parks in Bradford.

Other complete *Stigmaria* have been found in Britain and on the continent, but more importantly some groups of them have been discovered, providing evidence that they formed dense forests. The first was found in 1873 in Shef-

The 'Fossil Grove' in Victoria Park, Glasgow.

field; recent excavations charted their remains, but the site was then covered over to preserve them from further weathering. A more important group of eleven *Stigmaria* was uncovered in 1887 during excavations for a new park in Glasgow to commemorate Queen Victoria in her Jubilee year. The Scottish palaeobotanist Robert Kidston was involved in the excavation and played a crucial part in persuading Glasgow Council to construct a glass-roofed building to protect this now world-famous 'Fossil Grove.' This building has been open to the public ever since.

In the last few years *Stigmaria* and several erect stems have been uncovered at the site of the old Brymbo steelworks at Wrexham in North Wales and one *Stigmaria* has been removed and restored. It is more complete than those in the Glasgow Fossil Grove and unusually has a significant part of its erect stem. It is planned to preserve the exposure of Carboniferous rocks at Brymbo, and to use the reconstructed *Stigmaria* as the centrepiece of a display of Carboniferous plant fossils there.

## References

Cleal, C and B Thomas (2009). *An Introduction to Plant Fossils*. Cambridge University Press.

McGregor, M and J Walton (1972). *The Story of the Fossil Grove at Glasgow Public Parks and Botanical Gardens, Glasgow*, revised edition. Glasgow: Glasgow DC Parks Department.

Thomas, B and L Seyfullah (2015). *Stigmaria Brongniart*: a new specimen from Duckmantian (Lower Pennsylvanian) Brymbo (Wrexham, North Wales) together with a review of known casts and how they were preserved. *Geological Magazine* **152**, 871–901.

Williamson, W (1887). A monograph on the morphology and histology of *Stigmaria ficoides*. *Palaeontographical Society (Monographs)* **40**, 1–62.

# Gods and monsters

**Andrew Taylor**

Ecclesiastically significant geological sites are rare and I have had the good fortune to lead petroleum geology field trips to two of them. One kick-started Portugal's tourism in the eighteenth century and the other was used to support a creationist's world view.

On the western edge of Portugal, where Europe runs out of land, steeply dipping cliffs of Upper Jurassic limestones form the wind-sculpted headland of Cabo Espichel. It was here in 1410 that the Virgin Mary is reported to have arrived in Portugal, then ascended the vertiginous rockface on the back of a mule. This spot is marked by the fifteenth-century Ermida da Memória Chapel. Here the scene of arrival, including joyous angels, was captured on *azueljo* tiles in 1743. There are two sources of evidence for this event: the local fisherman that described the Virgin's progress, and the beautiful trackways that, due to tilting of the strata, appear from beneath the waves and can be traced to the cliff top. This is no donkey funicular, but one of the first sites to show herding behaviour in sauropods with parallel trackways of 10 individuals that walked at the same speed, although one of them had a pronounced limp (Lockley et al. 1994). Maybe this animal had a heavier load. Word spread through Europe about the significance of the *Nossa Senhora da Pedra da Mua* (our lady of the mule tracks) and this site became one of pilgrimage. It was Portugal's top tourist attraction until the town of Fátima gained fame in 1917 after a series of visitations by the virgin, above a tree, to three child shepherds. Coincidently, there is also a world-class sauropod trackway preserved in this village. However, there are no mule myths surrounding this apparition, and it is therefore not part of Fátima's religious experience.

In the UK the Cheshire Basin is dotted with outcrops and building stone quarries which have yielded excellent trackways from the Middle Triassic Sherwood Sandstone Group. These are so well preserved that associated sun-cracks, rain marks, and underfoot skin impressions can still be seen. The track resembles handprints with segmented fingers and outward-facing thumbs which were first named, in 1833 from German Buntsandstein material, as *Chirotherium* which means 'hand beast' (Seilacher 2007). Returning to Cheshire in 1857, material from Storeton quarry was used to construct the Christ Church, Higher Bebing-

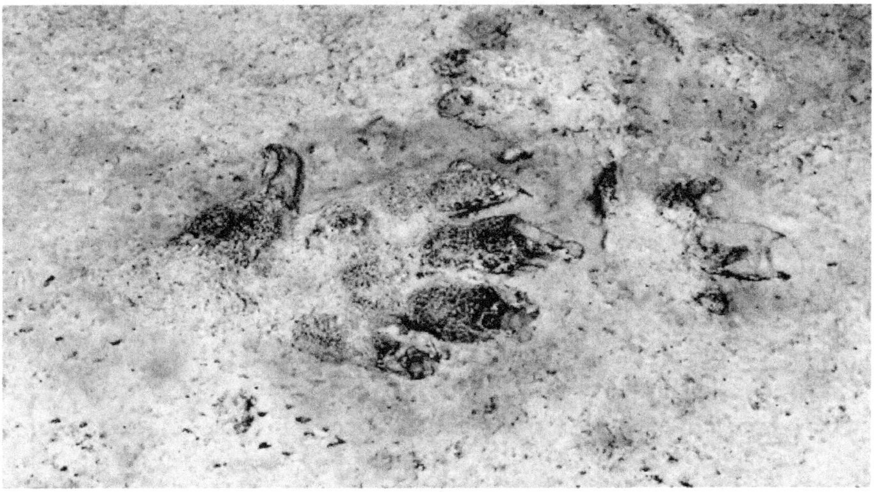

Close-up of the hand-like *Cheirotherium* pes and manus prints.

ton, where tracks span the porch. Considered by the locals to be the footprints of the devil, their evil powers were only neutralized by the Bishop of Chester consecrating the ground. More material went on public display (seen today in the Natural History Museum's main hall) and was the subject of study by both Richard Owen (1842) and Charles Lyell (1855). Proponents of creationism claimed them as proof that the world was geologically young and that if human imprints can be seen in stone we could not have evolved, like Charles Darwin suggested, from apes. However, the creationists did not look carefully at the material where both pes and manus prints are seen; some have scale impressions and the facing direction of the 'thumbs' would indicate that the 'hands' are on the wrong way round. This would indeed be a sign of the devil's work. Today, however, they are considered to be tetrapod rauisuchid reptile traces (Haderer 2001).

So there we have it, these ancient prints have been attributed to acts of God or the work of the devil and there is still a level of confusion about which body parts have been cast. They also started European religious tourism which is still thriving today, and perhaps could be summed up as 'virgin on the ridiculous'.

## References


Haderer, F (2001). Neues vom Handtier. *Fossilien* **3**, 172–174.

Lockley, M, C Meyer, and V Santos (1994). Trackway evidence for a herd of juvenile sauropods from the Late Jurassic of Portugal. *Gaia* **10**, 27–35.

Lyell, C (1855). *Manual of Elementary Geology*. London, 512 p.

Owen, R (1842). Description of parts of the skeleton and teeth of five species of the genus *Labyrinthodon* with remarks on the probable identity of the *Cheirotherium* with this genus of extinct Batrachians. *The Geological Society of London* **6**, 15–543.

Seilacher, A (2007). *Trace Fossil Analysis*. Springer, 226 p.

# Home-grown fossils

## Alex Cullum

It's not that I haven't found good fossils, or that they aren't there to find these days. It could be a little bit about conservation or maybe the fact that I'm getting old. OK, so I'm not sure quite why, but making fossils has become my latest obsession.

My fossils are not doppelganger Moroccan forgeries, I'm talking about machine-carved solid oak copies, meticulously sculpted from images of museum specimens. To be honest I didn't think it would be possible, but I've done it, so you could, too. I'm going to tell you how:

First you'll need a 3D milling machine. Not a 3D printer — those extrude liquefied materials which then set to form a model in plastic, concrete, steel, chocolate, etc. A 3D milling machine is a router or drill attached to a motorized frame which can move it in *x*, *y*, and *z* dimensions controlled by a computer. You provide a 3D model in software, and a block of wood, metal, plastic, or rock can then be cut with sub-millimetre accuracy to sculpt it.

Milling machines have until recently been extremely expensive, but in 2015 a company called Inventables released a kit in 1001 pieces at an affordable price. With YouTube assembly instructions even the electronics was easy enough.

Next you'll need software to build a 3D model of the thing you want to make. I use software from AutoDesk who very generously allow hobbyists to use some of their professional software for free. AutoDesk Fusion 360 is a perfect platform for creating and editing in 3D, it also has a Computer Aided Milling package which calculates the milling path, i.e. the sequence of 3D coordinates to send to the Arduio board that controls the milling machine.

My first 3D milled fossil was an ammonite which I drew completely within Fusion 360 but it took me ages. Complaining about this in the online forums, I was directed to another AutoDesk product called Remake which magically transforms multiple images from a normal camera into 3D models. Impossible you might think — I certainly did until I tried it.

Then the issue became finding some brilliant specimens to take pictures of. At this point the British Geological Survey saved the day. They are currently in the process of building an archive of 3D models from their extensive collections. It was really just a matter of picking which amazing specimen to try first.

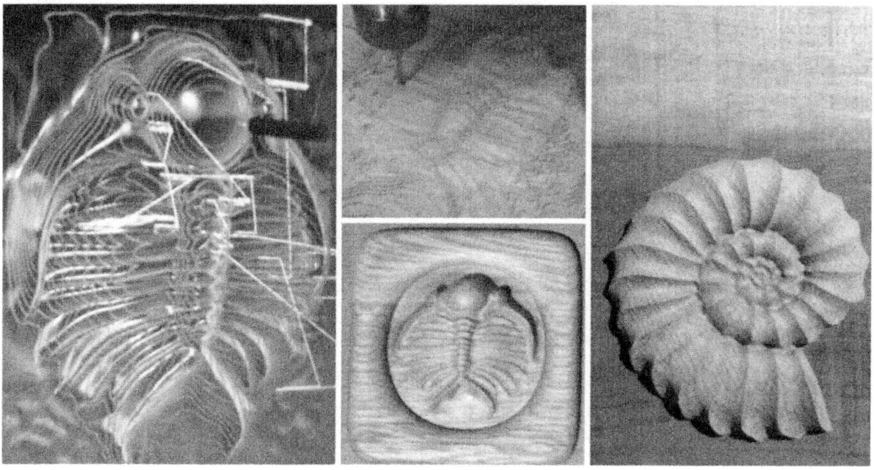

*Deiphon barrandei* model in Fusion 360 showing the drilling path lines. *Deiphon barrandei* in production and completed in oak wood (length ca. 15 cm). Ammonite in oak on a cherry background (ca. 25 cm diameter).

I eventually chose *Deiphon barrandei*, a trilobite from the Silurian.

Getting the files into Remake was easy enough. It's good for preparing the 3D model, cutting it down to just the parts I want, but getting these into Fusion 360 is harder. Most 3D files are based upon hundreds of thousands of triangles, forming a mesh. Fusion 360 uses squares and significantly limits their number. Nonetheless the required editing is still quicker and easier than drawing from scratch.

When the model is finished, you post-produce the milling path to a file called G-code, load that into Universal G-code sender, and it's time to get carving.

Did I make it sound too easy? Maybe a little. One catch is that the 'blank' or block of material from which you mill your fossil needs to be part of the model. So you need to be able to produce a blank to precise dimensions, which is not always as easy as it might sound. Then you need to anchor this in the machine. Again, it's not so easy to clamp something down, when that something is about to be cut away in 3D.

Although the milling process is noisy, dusty, and takes many hours, it's lots of fun — but not quite as exciting as finding real fossils.

## Further information

Arduino: *www.arduino.cc*

Autodesk: *www.autodesk.com*

BGS 3D archive: *www.3d-fossils.ac.uk*

Fusion 360: *www.autodesk.com/products/fusion-360/students-teachers-educators*

Inventables / X-Carve: *www.inventables.com*

Remake: *remake.autodesk.com*

Universal-G-Code: *github.com/winder/Universal-G-Code-Sender*

# How kangaroos got their bounce

## Benjamin P Kear

Kangaroos are an Australian icon, and one of the most unusual mammal radiations ever to have evolved. But when did these bizarre creatures first appear? How and why did this happen? These questions are being answered by fascinating new discoveries. But to unravel this story we must first understand what a kangaroo is.

Kangaroos, and other modern Australasian marsupials, have an evolutionary record stretching back more than 50 Ma. Yet their origins can currently only be inferred from the DNA of living species. This has confirmed the long-held view that kangaroos and their relatives — formally dubbed Macropodiformes — share a common ancestor with the arboreal brush-tailed possums, or Phalangeriformes. Moreover, by using gene mutation rates to estimate a molecular clock, a time constraint can be established for the initial 'bounce' down from the trees, and a clue derived as to when the ground-based kangaroo hop might have first developed.

Unfortunately, the earliest macropodiform fossils are unclear as to how this occurred. Indeed, the 'oldest kangaroos' such as *Palaeopotorous priscus*, are identifiable only from a few teeth. Nevertheless, these permit reconstruction of a superficially possum-like forest-dwelling animal that lived during the late Oligocene about 25 Ma. By 24–23 Ma, macropodiforms had diversified into archaic fanged 'kangaroos' called balbarids, and a disparate range of other more advanced macropodiforms epitomized by *Ngamaroo archeri*. This is the first archetypal 'kangaroo' known from an articulated skeleton, which possessed elongate hind limbs, an interlocking hinged-ankle, and functionally monodactyl foot with greatly enlarged fourth toe. These traits all represent essential equipment for hopping, and betray a close relationship between *Ngamaroo* and living kangaroos and rat-kangaroos — collectively termed Macropodoidea. Today the most genetically divergent macropodoid is the Musky rat-kangaroo, or *Hypsiprymnodon moschatus*, a rabbit-sized omnivore (living macropodoids are otherwise herbivorous) that does not hop, but rather uses a possum-like quadrupedal bounding gait. The skeleton of *Hypsiprymnodon* accordingly retains a swiveling ankle and small fifth toe, features that are typically lost in macropodoids. In fact, these peculiar characteristics occur in only one other macropodiform group — the extinct Balbaridae.

Balbarids were first recognized from dental remains that are 15 million years old (Middle Miocene). However, a spectacularly complete skeleton of Early Miocene (23 Ma) age, named *Nambaroo gillespieae*, has established Balbaridae as a completely separate macropodiform radiation that bounded, probably climbed, and perhaps even ate meat based on their close relationship with Propleopinae — these were the famous 'carnivorous kangaroos' that constituted a probable balbarid sister lineage. It is the extinction of balbarids that offers the most important insight into possibly why kangaroos started to hop.

And the answer is… because they were already hopping.

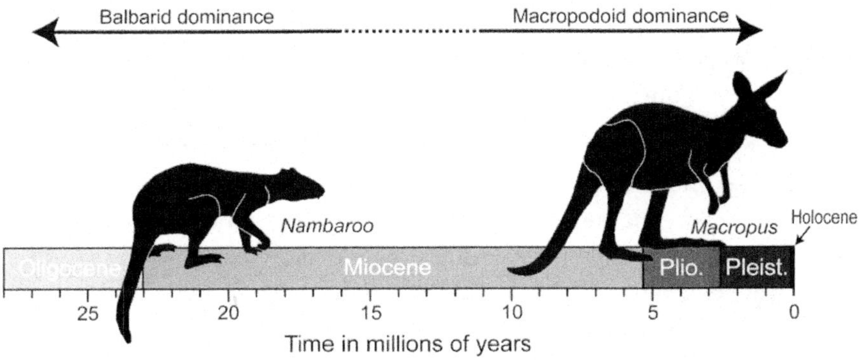


All macropodiforms, even balbarids, possessed skeletal modifications for a hopping gait. It was just whether the fundamental synchronous movement of the hind limbs took place in alternation with the forelimbs during a quadrupedal bound, or bipedally in conjunction with elevation of the torso. The key event occurred somewhere around 15–10 Ma when the forested world of the balbarids gave way to dry grasslands that favoured more upright hopping macropodoids. Their most crucial specializations included high crowned molar teeth capable of grinding tough plant matter, and enlarged body size to accommodate a voluminous fermentative gut. Increasing aridity after 10 Ma propagated savannas and deserts, which again selected for macropodoids. Thanks to these profound climatic alterations, therefore, their descendants — including the quintessential grey kangaroo *Macropus giganteus* — have come to dominate almost all terrestrial herbivore niches across Australia and New Guinea. The much-celebrated kangaroo hop was simply inherited from ancient progenitors such as *Ngamaroo*, and then eco-morphologically fine-tuned to produce an adaptable locomotory apparatus that has since complimented the persistent success of macropodoids in the face of changing environments.

# Impossible frogs in the Deccan Traps

**Michael Oates**

This story started when I was about to transfer to Mumbai. As a palaeontologist, reading up on the geology was depressing. I was moving to the middle of the half-million-square-kilometre Deccan basalt lava plateau. The discovery in the mid-nineteenth century of small fossil frogs, in some ashy, shaly sediments between lavas in Mumbai was dismissed as an impossible dream, but not entirely forgotten.

For over a year, I thought little more of it, especially when discovering that the best place to find them was in a worked-out quarry now completely covered by a slum. Believe me, you don't want to go digging around in a Mumbai slum environment if you value your health. I then learned that St Xavier's College, which teaches geology, had a decent collection of these frogs. They had been open to public viewing a week before it came to my notice; it was too late to go. I talked my way into an introduction with the college professor, through an ex-student I had met through work. She obliged and I had time on a trip to the UK to go into the field and collect a decent amount of British Jurassic material from the type locality of one of the Jurassic stages. I took the whole lot over to India as a present for the college, with an inkling that they might reciprocate — but this was asking a bit too much.

In the event, the departmental head, Dr Hrikishesh Samant, was very impressed with the collection. But when he learned that I was a graduate of University College London, he announced that he had been collaborating with a professor in the zoology department who dearly wanted to look at some frog material to determine if it was possible to see it using a CAT scanner. He asked if I would like one of the frogs, which he himself had collected some 30 years back, during building work at another site. The offer was conditional on me hand delivering it to Professor Evans at UCL in the UK for the tests. I could hardly believe my luck, and gladly agreed. He told me that the location at the base of Malabar Hill in the heart of the old city was long since gone, but let me know the whereabouts anyway. A week later, while passing this spot, I asked my driver to locate the 'Bombay Garage', behind which the frogs had been found. We found, at the base of a steep lava scarp, the shales still cropping out amid a lot of rubbish behind illegal hutments. I collected a few pieces and, later that day at home, found a

*Of particular scientific interest is… how such a small, environmentally sensitive group of creatures survived one of the biggest extinction events on Earth.*



couple of tiny leg bones in the material. Encouraged by this, I returned to gather more rock. Considering where it came from, it seemed wise to wash each piece carefully. From one large chunk fell a few flakes which I almost discarded but decided to look at under my binocular microscope just to be sure. My eyes almost popped out of their sockets. There lay an exquisite frog skeleton, unseen by any creature since the Early Palaeocene 60 million years ago.

Left: *Rana pusilla* (Owen 1847) (syn. *Indrobatrachus pusillus*), Palaeocene intra-Trappean shales, Mumbai, India. Scale: about 3 cm wide. Right: The Palaeocene intra-Trappean shales, Malabar Hill, Mumbai, India.

My personal stretch target had been met. But what is of particular scientific interest is not just the position of these frogs in the evolutionary sequence, but how such a small, environmentally sensitive group of creatures survived not only one of the biggest extinction events on Earth, but also the challenging conditions of frequent local volcanic outbursts. History records that turtles and crocodile bones also occur in these deposits, along with abundant ostracod and rare freshwater mollusc shells. The 25 kg of material I imported to Stavanger to scrutinize during the long winter nights may not yield any of these, but so far I have found the remains of a further 20 individual frogs.

# In defence of the shoehorn

## Roger Mason

I have recounted the events of my discovery of *Charnia masoni* in Charnwood Forest near Leicester many times on radio, TV, in lectures, and in print. The story has acquired additions and variations, some true, some false, and has a life of its own on the internet. The key event was Trevor Ford's publication of the identification of this Precambrian fossil in 1958. Ediacaran fossils were first seen and recognized in 1946 by Reginald Sprigg, an Australian geologist who was investigating a proposed dam site in the Flinders Ranges north of Adelaide, South Australia. He had to leave before he could make notes because of a scorpion bite and he did not come back until a mineral prospecting trip with the prime minister of South Australia in 1948. A cartoon in his autobiography shows him on this occasion with a fossil that appears to be *Charnia* under his right hand. He submitted a manuscript describing the fossils to *Nature*, but it was rejected.

Trevor Ford's publication of the Charnwood discovery stimulated a flurry of activity in Australia. *Nature* accepted a paper by Martin Glaessner in 1959 and the fossils were generally accepted as Precambrian. Sprigg called them 'jellyfish' and Glaessner identified them as Coelenterata (Cnidaria in modern terminology) following the prevailing belief of the time that cnidarians were the most primitive phylum of multicelled animals. Ediacaran fossils were subsequently found in all continents except Antarctica and in a wide variety of sedimentary rocks and palaeoenvironments. The majority of the fossils are impressions on bedding surfaces, like *Charnia*.

Adolf Seilacher stirred things up in 1992 by arguing that the Ediacarans were not cnidarians but an extinct group of organisms with body plans unlike animals, plants or fungi, initiating re-examination of the fossils and causing debate that continues today. This later research convinces me that *Charnia* and most other Ediacarans were not cnidarians, but there seem to be as many theories about what they actually were and how they lived as there are research teams studying them. They displayed a regular pattern of evolutionary change between 635–541 Ma before the Cambrian period, and in 2004 this persuaded the International Union of Geological Sciences (IUGS) to declare formally an Ediacaran period, the first new geological period to be recognized in 120 years.

The Australian government issued a special set of commemorative stamps and the Ediacara district has been designated a World Geopark.

The Ediacaran period was a dramatic time in Earth history. It followed a world-wide glaciation when most of our planet's land and oceans were covered in ice during a 'Snowball Earth' period, and was succeeded by the explosion of multicellular living organisms at the beginning of the Cambrian. There were dramatic changes in the chemistry of the oceans and atmosphere. How these events relate to the appearance and disappearance of the Ediacarans continues to be a lively topic of research and discussion.

I encountered Ediacarans again when I was doing geological research near the Chinese city of Wuhan. The fossil *Paracharnia dengyingensis* had been discovered in Late Proterozoic rocks beside the Yangtze River in a survey ahead of the completion of the Gezhouba Dam near Yichang in 1983. Glaessner examined the fossil and his Chinese collaborator made a fanciful reconstruction of it as a cnidarian. The location is now beneath a reservoir but Ediacarans have recently been found at several other localities in the uppermost Proterozoic of the Three Gorges Region, accompanied by acritarch microfossils and millimetre-scale macrofossils.

The story of the Ediacarans highlights the difficulty of interpreting very ancient fossils. In his 1989 book *Wonderful Life* Stephen Jay Gould castigated earlier misinterpretations of Cambrian Burgess Shale fossils of western Canada, labelling it 'shoehorning' — meaning fitting fossil organisms into known biological categories despite significant differences. Reinterpretation of the Ediacara biota since Gould died in 2001 demolishes his argument that Charles Walcott, the discoverer of the Burgess Shale fossils, shoehorned them because he was distracted by his duties as Secretary of the Smithsonian Institution in Washington. Despite their different responsibilities, Sprigg, Glaessner, and others shoehorned the Ediacarans into the phylum Cnidaria. Shoehorning is not disreputable science but a reasonable attempt by pioneer palaeontologists to understand astonishing new discoveries without the benefit of Gould's hindsight.


## References

Ford, T (1958). Pre-Cambrian fossils from Charnwood Forest. *Proceedings of the Yorkshire Geological Society* **31**, 211–17.

Glaessner, M (1949). Fossil Coelenterata from Australia, Africa and England. *Nature* **183**, 1472–73.

Gould, S J (1989). *Wonderful Life*, New York: Norton.

Seilacher, A (1992). Vendobionta and Psammocorallia: lost constructions of Precambrian evolution. *Journal of the Geological Society*, London **149**, 607–613.

# In search of the Balearian mouse goat

## Alun Williams

I'm happy to admit it — I'm a bit of a nerd when it comes to geology. Not a 'know which beds and fossils define every formation' nerd, but the sort of nerd who likes to look at rocks in his time off, despite the fact that I'm a professional geologist. And I particularly enjoy doing this when I'm on holiday, as what better opportunity is there to see a random selection of new rocks?

I prepare for my holidays as if I'm about to undertake a minor piece of fieldwork. A few weeks before I travel, I'll start to collect publications on the region, which I'll read and narrow down to a few key papers to print out and bring along with me. Initially, I'll aim to get a broad understanding of the geology of the region I'll be visiting, but as I dig deeper, I'll try to identify particular outcrops that might be worth a visit. And then I work out how much time to negotiate away from my family, or whether I should propose a visit to some family-friendly feature nearby.

And so it was with Mallorca — a small slab of Tethys that's been thrust above the azure waters of the Mediterranean, with a huge variety of accessible rock exposures ranging in age from the Triassic to the present. Mallorca has many geological highlights, but the real gem for me was located in the southwest of the island.

Near the town of Cala Figuera a series of Pleistocene aeolianites, wind-blown dunes composed largely of carbonate, are exposed as sea-cliffs. These rocks lithify rapidly, preserving all manner of sedimentary features, including a wide variety of terrestrial trace fossils. And these particular dunes contained the tracks of an extinct animal called *Myotragus balearicus*, a name that translates as the Balearian mouse goat. What geologist wouldn't be excited by this?

*Myotragus balearicus* is a goat that became isolated on the Balearic archipelago, and evolved into a dwarf form in order to survive in the arid rocky landscape. Their ancestors found their way to Mallorca during the Messinian salinity crisis, around six million years ago, when the Mediterranean Sea evaporated almost entirely. They were trapped by the subsequent rapid sea-level rise that set in train a remarkable series of evolutionary adaptations, following a process known as island dwarfism. This is commonly seen in isolated populations, and involves

Left: Beautifully preserved Pleistocene aeolianites at S'Estret des Temps, southern Mallorca. Right: Trackways of the Balearian mouse goat, *Myotragus balearicus* (€2 coin for scale).

large animals becoming smaller through successive generations. An inverse process called island gigantism is equally common, and resulted in a giant dormouse living on Mallorca at the same time as *Myotragus balearicus*, as well as a giant rabbit on the neighbouring island of Menorca.

Even more remarkably, the bone structure of *Myotragus balearicus* shows similar growth patterns to those of crocodiles and large lizards, leading to a proposal in 2007 that the mouse goat may have been cold-blooded, a useful adaptation in animals that have to survive long periods without food. The reason for their eventual demise around 4000 years ago is a topic of some debate, although it seems to have coincided roughly with the arrival of humans to the island. This is unlikely to be a coincidence, and it has been proposed that attempts were made to domesticate the mouse goat, although apparently without success.

Finally, these strange animals were first discovered by Dorothea Bate, a remarkable woman from west Wales, who became one of the founders of the field of archaeozoology despite having received little formal education. But her story deserves a telling of its own.

## References

Fornos, J, R Bromley, L Clemmensen, and A Rodriguez-Perea (2002). Tracks and trackways of *Myotragus balearicus* Bate (Artiodactyla, Caprinae). In: Pleistocene aeolianites from Mallorca (Balearic Islands, Western Mediterranean). *Palaeogeography, Palaeoclimatology, Palaeoecology* **180**: 277–313.

# Integrated Silurian biostratigraphy

**David Loydell**

Graptolites are astonishingly useful fossils. For reasons that we can never know, they evolved very rapidly, spreading quickly from where they first appeared and colonizing the Silurian oceans in great numbers. They have the major advantage over many microfossil groups that they are almost never reworked from older strata (except within large conglomeratic clasts). They also have the ability to survive weathering, such that they can be collected and identified from heavily weathered surface exposures where all traces of microfossils have been obliterated by oxidation. As a result, graptolite data can be available through Palaeozoic sections where little else is present. This provides constraints on the timing of geochemical and geophysical data which otherwise are often no more than intriguing wiggly lines with no evidence as to the age of, for example, a particular carbon isotope excursion or interval of high gamma-ray values. Graptolites have enabled recognition of major tectonic features such as accretionary prisms (e.g. in the Southern Uplands), dated innumerable hydrocarbon source rocks, made possible the accurate geological mapping of extensive areas on many continents, and provided the framework for understanding the growth of submarine fans, such as in the Welsh Basin.

So, having written a manifesto on behalf of the Graptolite Party, it's now time to explain why one cannot live by graptolites alone and why integrated biostratigraphy has been the focus of research efforts in recent years.

*Spirograptus guerichi* from El Pintado reservoir, Seville province, Spain. The graptolite is 12 mm long.

*[Graptolites] have the major advantage over many microfossil groups that they are almost never reworked from older strata.*



It is clear that graptolites favoured certain conditions found in outer shelf and deeper waters; their abundance and diversity declined dramatically in shallow water. As sedimentary rocks from shallow marine environments contain very few or no graptolite fossils, dating those rocks requires conodonts and/or chitinozoans.

By the mid 1990s there were biozonations available for deep-water sections based upon graptolites and for shallow-water sections based on conodonts, but very little idea as to how the biozonations might correlate with each other. By comparison with studies on graptolites and conodonts, those on Silurian chitinozoans were in a state of relative infancy, but it was clear that these fossils had considerable biostratigraphical potential and occurred in a wide range of environmental settings. The search was then on for those sections that were just deep enough for there to be a decent graptolite record, also contained chitinozoans, and were sufficiently shallow for conodonts to occur in sufficient numbers to recognize their biozones. Such sections, mostly from the palaeocontinent Baltica, have enabled the integration of the conodont, chitinozoan, and graptolite biozonations based on fossils present within a single sample or from adjacent samples within a single section. As a result we can now correlate with much greater precision from shallow into deep water and have the opportunity for much more refined interpretations of the major biological and geological events affecting the Silurian world. For example, we now know that graptolite and conodont extinctions were not synchronous: that graptolite extinctions coincided with the onset of falling sea level and positive carbon isotope excursions whilst conodont extinctions preceded them.

Needless to say, much more work remains to be done — many parts of the world remain largely unknown with regard to their graptolite, conodont and/or chitinozoan biostratigraphy. Considering that this work provides the key dating constraints upon which all other geological studies depend, this may seem surprising. The problem, however, is not the lack of work to do but the few people trained to do it.

# Interview with the *Triceratops*

**John Scannella**

As I write this, there is a *Triceratops* resting in the back of one of the Museum of the Rockies' pickup trucks. The truck is parked beneath the stars of eastern Montana and in a few days it will be on its way back to the museum. Once at the museum, preparators will begin the process of removing the layers of burlap and plaster that have been used to wrap the fossils for safe transport, and will carefully chip away at the rock that has encased this animal for over 65 million years. It will be the culmination of a lot of work to excavate this horned giant, but once the fossils have been prepared this *Triceratops* will be able to speak to us, to teach us something about itself and its world.

*Triceratops* is the most commonly recovered dinosaur in the uppermost Cretaceous Hell Creek Formation of Montana (Horner et al. 2011). In some places, *Triceratops* are literally spilling out of the hillsides. Many dinosaur species are represented by small sample sizes, maybe one or a few specimens; the large number of *Triceratops* that have been collected is one of the things that makes this animal vital to our understanding of the Cretaceous. Large sample sizes provide opportunities to explore variation between individuals. A decade ago, after the collection and study of numerous specimens of various sizes, it was discovered that the skull of *Triceratops* changed dramatically as it aged (Horner & Goodwin 2006). Differences between specimens that were previously suggested to indicate distinct species were found to instead represent growth changes, in much the same way that many animals today (including humans) can look quite different as they mature.

We can learn even more by studying the layers of rocks that fossils are found in. When we were graduate students at Montana State University, Denver Fowler (see *Dinosaur evolution: it's about time*) and I spent several summers in the Hell Creek Formation, trying to determine where, in time, many *Triceratops* had been found. We spent many hot days climbing up and down hills in the badlands of Montana, measuring the positions of *Triceratops* quarries (Scannella & Fowler 2014). Was the animal found low in the formation? Or in the upper layers, near the end of the Age of Dinosaurs? In a way, we were following *Triceratops* through time, because as you climb up from the bottom of the Hell Creek Formation to the top, you are climbing up through time. And,

*As more* Triceratops *are collected, we can only learn more; new hypotheses will be formulated and some will be falsified.*

interestingly, we found that the *Triceratops* we encountered appeared to change along the way.

Currently, there are two recognized species of *Triceratops*: *T. horridus* and *T. prorsus* (Forster 1996). *T. horridus* tends to have a short nasal horn whereas *T. prorsus* has a more elongate nasal horn. Could this actually be a case of differences between maturity stages, or perhaps sexual differences? By studying the position of each animal in the rocks, we see that *T. prorsus* is only found high in the formation and *T. horridus* is recovered from the lower layers. As you climb up through the Hell Creek Formation, the nasal horns you encounter tend to get longer. By examining the details of *Triceratops* from different levels of the formation back in the lab, it became apparent that several other features of the skull were changing over the course of the latest Cretaceous. This suggests that *T. horridus* evolved into *T. prorsus* (Scannella et al. 2014).

Every *Triceratops* has something to tell us. Perhaps about growth, evolution, behaviour, or the role they played in the world at the end of the Cretaceous period. As more *Triceratops* are collected, we can only learn more; new hypotheses will be formulated and some will be falsified. That's why, as I write this, I'm excited that there's a new *Triceratops* waiting to be brought back to the lab, freed from the rock, and given the opportunity to speak to us.

## References

Forster, C (1996). Species resolution in *Triceratops*: cladistic and morphometric approaches. *Journal of Vertebrate Paleontology* **16** (2): 259–270.

Horner, J and M Goodwin (2006). Major cranial changes during *Triceratops* ontogeny. *Proceedings of the Royal Society of London* B. **273**: 2757–61.

Horner, J, M Goodwin, and N Myhrvold (2011). Dinosaur Census Reveals Abundant *Tyrannosaurus* and Rare Ontogenetic Stages in the Upper Cretaceous Hell Creek Formation (Maastrichtian), Montana, USA. *PLoS ONE* **6**:e16574.

Scannella J, and D Fowler (2014). A stratigraphic survey of *Triceratops* localities in the Hell Creek Formation, northeastern Montana (2006–2010), 313–332. In: *Through the End of the Cretaceous in the Type Locality of the Hell Creek Formation in Montana and Adjacent Areas*, G Wilson, W Clemens, J Horner, and J Hartman (eds.) Geological Society of America Special Paper, Boulder, Colorado.

Scannella, J, D Fowler, M Goodwin, and J Horner (2014). Evolutionary trends in *Triceratops* from the Hell Creek Formation, Montana. *Proceedings of the National Academy of Sciences* **111**: 10245–250.

# Keep an open mind

**Robert Scott**

The discovery of cactus-like moulds in uppermost Lower Cretaceous marine limestone was the result of curiosity and an open mind. A graduate student noted the moulds as he explored limestone beds in well-exposed road cuts. I casually dismissed them as drill holes. But the sight nagged me until new observations exposed their hexagonal cross sections with angular facets bearing small circular rings on the crests. Finally someone suggested that the shape looked like a pet cactus.

Erect, well-preserved moulds of a fossil stalk-like organism are buried in marine limestone about 100 Ma old on the Comanche Shelf in West Texas. The upper Albian age is defined by ammonites in the lower beds and by caprinid rudists and benthic foraminifers in the upper part of the formation.

The moulds are up to 1 m tall and about 6 cm in diameter and have hexagonal cross sections with vertical ribs bearing aureoles separated by grooves or flutes. The mould walls are composed of six pairs of angular ridges and grooves, evenly spaced at an angle of 60° to each other. The ridge crests are dotted by paired ringed depressions with small central nodes; these negative impressions indicate the presence of oval rings with small depressions inside each ring. Certain modern cacti have similar structures, which are tubercles and areoles from which spines grow. As yet spines have not been recognized; these fine organic structures may have separated from the stems and become part of the sediment. They may be preserved as small organic particles or as moulds that would be similar to pores. The diameter of the stalk is the same at its base as at its top indicating that the organism did not expand as it grew upward. The vertical orientation of the stems indicates that they were buried slowly in low-energy conditions. The bases of the stalks are rooted in red clay palaeosols that separate beds deposited by cyclical flooding–receding marine waters. Carbon isotope chemostratigraphy suggests that the disconformable bedding surfaces were exposed to the atmosphere.

The evolutionary origin of the family Cactacea is obscured by the near absence of a fossil record of either pollen or plant parts (Arakaki et al. 2011). Pollen of modern cacti is small and fragile, thus unlikely to be preserved. One hypothesis proposes that cacti evolved in the New World in latest Cretaceous after the

separation of Gondwana because their modern distribution is restricted to South America and southern parts of North America (Anderson 2001). However, multiple line of evidence suggest that cacti evolved in the latter part of the Palaeogene period in South America. Contemporary hybrid phylogenomic methods support this, suggesting that cacti evolved from their ancestors about 35 million years ago in the latest Eocene–earliest Oligocene in the New World (Arakaki et al. 2011). Since then, modern cactus groups diversified beginning about 10–5 million years ago as $C_4$ grasslands expanded across uplifted plains. Regional arid environments spread across continents as atmospheric $CO_2$ and temperatures decreased. Discovery of cactus-like stems in strata at about 100 Ma extends the geologic range by up to 60 million years and challenges the current phylogenomic methods.

The fossil moulds represent a new genus and species, which is characterized by its erect cylindrical columnar stem with a hexagonal cross section and vertical ribs bearing aureoles separated by grooves or flutes. The stem morphology is comparable to others in the subfamily Cactoidea, which is characterized by cortical vascular bundles forming 3D networks in the cortex (Nyffeler & Eggli 2010). Unfortunately the key internal structure of this Cretaceous fossil is unknown. This subfamily includes the tribe Cereeae as the molecularly most primitive cactus. This Albian cactus is the oldest and therefore the most primitive fossil cactus known as yet.

Mould of a cactus-like fossil stem showing three faces and a flat base and the outline of pores on angular ridges (Scott et al. 2016). The scale bar is 7 cm.

## References

Anderson, E (2001). *The Cactus Family*. Timber Press, Pentland, Oregon.

Arakaki, M, P-A Christin, R Neyffeler, A Lendel, U Eggli, R Ogburn, E Sprigs, M Moore, and E Edwards (2011). Contemporaneous and recent radiations of the world's major succulent plant lineages. *PNAS* **108** (20), 8379–84.

Nyffler, R and U Eggli (2010). A farewell to dated ideas and concepts: Molecular phylogenetics and a revised supra generic classification of the Family Cactaceae. *Schumannia*, Bonn, **6**, 109–149.

Scott, R, X Wang, and Y Wang (2016). Late Albian cactus in shallow carbonate shelf, West Texas. *Carnet Geologie* **16** (12), 337–346.

# Life and the selfish universe

**Mike Whitaker**

I was having a coffee with a friend back in 1968. He placed two cups on the table, one full, one half-full. He pushed the half-filled one towards me.

"You're selfish Martin," I said.

"I know!" he said proudly. "Everything I do is selfish... the universe is selfish."

The idea that developed from this initial conversation shows how the universe, initiated by the massive release of energy referred to as the Big Bang, immediately became pre-occupied with energy management. An orderly synthesis of basic particles — atoms and molecular structures — provided an efficient way of storing energy. This organization slowed the decay of the universe and prevented it disappearing in a flash of light.

Further energy conservation is achieved by combining atomic structures to form molecules. Organic molecules, specifically the self-replicating DNA, represent a lower energy still, and introduce the important additional mechanism of lower energy conservation with a unique ability to employ energy from the sun. This places ourselves and all other living forms as a logical part of energy conservation within the universe. As far as the universe is concerned this is our only function and the only reason we exist.

The processes at work seem relentless and ruthless. Phenomena analogous with the human characteristic considered 'selfish'. In this context, imagine an evolving virus-like molecule, eventually deviating into RNA and then DNA. Now, unlike a normal virus, it has a replicating system capable of infinitely regenerating and modifying itself. Over time, this process built ever more sophisticated organic self-supporting hosts, protecting the 'virus' within the nuclei of the cells as DNA.

Continued survival demands the DNA to steer our way of thinking and consequently all our thoughts, emotional experience, and behaviour. This is achieved by a tripartite neural system, in part, which effectively motivates us to do the optimal thing for survival. Starting with the ability of the reptilian brain to signal the motion to flee or fight with the release of adrenalin, the

*The tripartite brain rewards problem solving and creative thinking, which are important survival factors.*



brain developed further lobes gradually increasing sensory data. To maintain an orderly input of this data a frontal lobe receives all the sensory data for a period of about 10 seconds. It appears to assess this period for survival and signals the reptilian brain, which assigns a proportionate amount of adrenalin and dopamine, providing us with an amount of anxiety or pleasure to stimulate the appropriate action. Failing to do the right thing for survival results in withdrawal of dopamine and we feel the consequent discomfort (guilt, frustration, anger). In addition, the tripartite brain rewards problem solving and creative thinking, which are important survival factors. Hence the pleasure we get from scientific discoveries, new musical structures, humour, and solving crossword puzzles. It appears that the strength of the signal may be strong or weak and could vary with each individual. For example, a strong signal to the reptilian brain can result in an excitable and more nervous, emotional individual. Similarly, a weak signal could produce a cooler individual, who is braver and less afraid. An extremely weak signal would not achieve any emotional response. Without experiencing elation or remorse, behaviour may tend to extremes attempting to generate any kind of emotion. (Psychopaths may represent this condition.)

In effect our DNA, via the tripartite brain, steers our every move and emotion by reward and withdrawal to ensure survival. But the DNA is only conforming to the physical laws of the universe. We carry out our lives via the DNA in a blind obligation to satisfy the precise energy requirements our particular chemicals require. Our daily life is steered with short bursts of happiness, tempered with withdrawal, just enough to engage us in the pursuits that ensure the survival and reproduction of the DNA.

So if anyone asks you, 'Who are you and why are you here?' perhaps the answer should be, 'I am the host to a parasitic virus and I'm here to conserve energy and help save the universe.'

# Little big mammals

**Jérôme Prieto**

I'm fascinated by the question of how terrestrial life evolved in relation to environment. My area of specialization is the study of small — really small — mammals. But what exactly are small mammals?

When I try to explain my profession to friends, they intuitively conclude that I work on ancient animals related to present-day rodents, bats, rabbits, and insectivores. And, yes, all these animals are small. Some of my friends regard these animals as cute, while others consider them dirty or disease-ridden.

Compared with the barrel-shaped South American capybara, a rodent that could weigh as much as an adult person, many rodent-like mammals would appear small. Even the giant capybara would be small compared to the fossil *Josephoartigasia*, a giant Pliocene–Early Pleistocene rodent that could grow up to 4 m long from nose to tail end.

With regard to the European Middle and Late Miocene faunas (i.e. ~16 to 6 Ma), the term 'small mammals' is traditionally used to refer to bats, lagomorphs, rodents, insectivores, and marsupials, the latter infraclass being poorly diversified, with species close to the opossums (not a kangaroo!). With but a few exceptions, the fossils of all these animals are found disarticulated, mostly highly fragmented, and the pieces are most often obtained by screen-washing of sediments followed by a fastidious process of picking under the microscope. Isolated teeth found in these fossil assemblages typically are no larger than 1–3 mm long, but still represent the most important fossil tools in the recognition of extinct small mammals.

A wide variety of dental features are used today to discriminate between species based on isolated teeth fossils. However, tooth size and morphology play a major role. Basic features of the chewing surface are important and each of the ridges (crests), peaks (cups), and valleys has its own scientific name, and their relative size, depth, connection, or simply presence or absence may be significant as classification criteria. As a result, scientific descriptions of fossil teeth for the layman must seem to be amalgamations of incomprehensible names bound together by verbs, and spiced up with adjectives, numbers, or percentage indications.

Small teeth will hardly ever be presented in major museum exhibitions and admired by large numbers of visitors. Rather, the beauty and importance of these fossils lies in their scientific value.

There are several basic tooth morphologies that permit safe recognition of certain groups of animals. For example, sharp cups are characteristic of bats and insectivores, while small molars with long transversal crests might be indicative of dormice. Furthermore, fine differences (e.g. the number and length of the crests) allow us to differentiate between species within one genus.

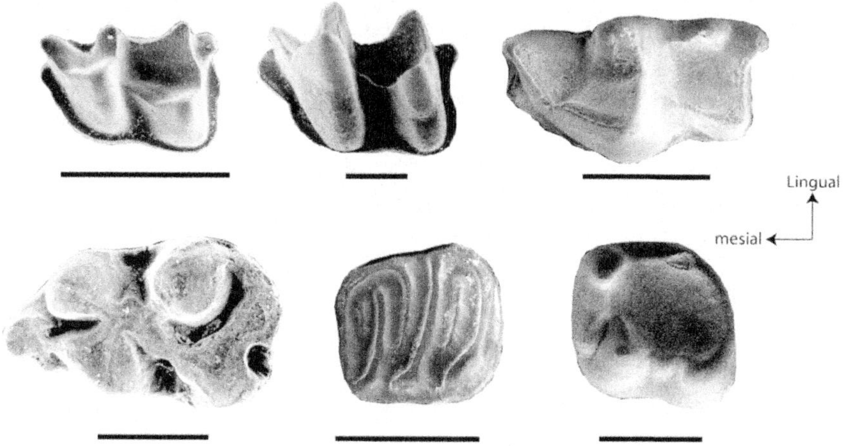


Morphologic patterns of some Middle Miocene European small mammals. All teeth are lower molars presented in the left orientation. From left to right, first line: bat (cf. *Myotis* sp.), insectivore (mole, *Desmanodon fluegeli*), and marsupial (*Amphiperatherium frequens*). Second line: cricetid rodent (*"Cricetodon" fandli*), dormouse (*Miodyromys biradiculus*), and squirrel (*Spermophilinus* sp.). Scale bars are 1 mm.

The age of a fossil fauna can be assessed based on teeth. In a peculiar geographic region, each species of animal is characterized by its first occurrence in the fossil record (oldest possible dating of the fauna), its disappearance from the record (youngest dating), as well as its evolutionary level (intermediate dating). If a given species shows size increase through time, it is possible to estimate an age based on the size of the teeth. Because small fossils are intricately linked with the sediments in which they are preserved, the dating also allows assessment of even larger-scale geological processes such as the development of mountain ranges, or even the closure of seaways. On the other hand, a comparison of the small mammal fossils from different geographical regions may provide insights into the migration of certain animals that, in turn, can be used to reconstruct climate changes through time.

These tiny fossils have a surprisingly big influence on our understanding of past life and ecosystems.

# Making sense of swamps

**Jon Noad**

I am sure that many geologists were ardent fossil hunters as children, and indeed may have been inspired to continue with a career in earth sciences as a result. At some point along the way many geoscientists come to realize that fossils are much more than stunning examples of nature's artwork; they are also valuable tools in interpreting both the age and sedimentary environment of the rocks in which they occur. When both body fossils and trace fossils are present, providing the opportunity to interpret and incorporate animal behaviour into the overall depositional model, it is a powerful combination. This is the story of one such find.

When you walk into a quarry you never know quite what to expect. When it is located in a remote corner of eastern Borneo, where your ancient geological map shows only undifferentiated Miocene deposits, it comes as a huge surprise when an outcropping sandstone bed reveals beautifully preserved fossil bird footprints. Piecing together who made them, and in what setting, opened up a new chapter of research in my work.

The small tridactyl prints lacked heel prints, suggesting that they were made by a shore bird similar to a plover, and the birds had even left a few 'prod' marks, where they had probed the sediment with their bills. Beyond this I needed more information.

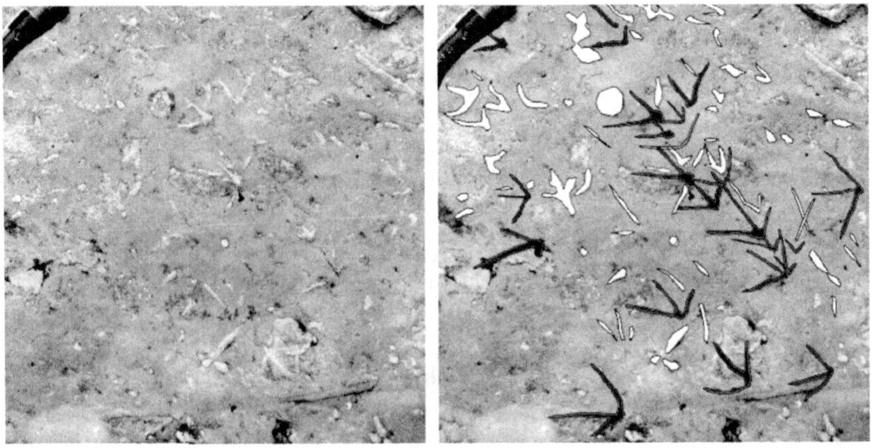

Fossil bird footprints from the quarry on the left, with my interpretation on the right.

Further examination of the sheet-like sandstone beds revealed more fossil trackways. There were a series of deep, paired, conical pits, which comparison with modern animal tracks suggested were made by a deer. Alongside these were lines of paired, teardrop-shaped impressions, characteristic of the footprints of some kind of crustacean, as well as casts of fossil leaves and fan-like traces, left by vegetation washed downstream. Was this a coastal channel of some kind? There were definitely a few narrow, steep-sided, channel-shaped sandstone bodies, but other sandstone beds were flat and wide.

I turned my attention to the intervening, thicker, mudstone beds. Fine-grained sediments typify lower energy deposits, and often provide better fossil preservation, while weathering will commonly expose body fossils which have either been cemented or have hard shells. The first thing to catch my eye was a series of rooted trees with buttresses. The woody material had been petrified by silica, preserving tree rings and even xylem. The preservation was so good that I could see where marine shipworms, actually boring teridinid bivalves, had penetrated the wood.

So was this a flooded forest or a coastal swamp? A few moments later I located the final piece of the puzzle, a beautifully preserved fossil 'yabby', or mangrove lobster, encased in an ironstone matrix. In modern mangroves these animals are capable of processing tons of sediment daily, making mounds up to a metre in height. Their presence, as hand-sized fossils, confirmed the identity of our crustacean track maker, as well as the depositional environment in which our plovers were foraging — the lobsters occur only in mangroves. The trackways proved both that these beasts had walked across the sandy flats, and that their body fossils were *in situ*, not washed into this ancient Miocene mangrove.

Integrating sedimentological data with the ichnological (trace fossil) and palaeontological (body fossil) evidence allowed me to come up with a unique depositional interpretation for these sediments. The sandstone channels and muddy flats rich in mangrove lobster tracks, populated by buttressed trees bored by marine bivalves, could only be explained as the remains of ancient mangrove swamps. These have rarely been identified in the fossil record, so these new recognition criteria should help to alleviate this situation, and reinforce the need to integrate all the available fossil and sedimentary data in order to derive the most appropriate depositional model.

# Micro-dung and its uses

**Wyn Hughes**

Organisms of all sizes eat and inevitably have to expel undigested matter. The study of microfossils normally deals with the identification of the fossilized remains of hard-shelled organisms, but some microfossils represent the products of their host. The identification of fossilized excrement is termed parataxonomy, and their identification and species determination relies purely on shape and, where visible, internal features.

## *Favreina*

Thin sections of limestones of Late Triassic age and younger, often show subrectangular to rounded dark shapes that display internal chain-like patterns or parallel streaks. These shapes are fossilized cylindrical faecal pellets called microcoproliths, and were produced by decapod crustaceans. They are typically around 1.5 mm long and 0.75 mm in diameter. In gross appearance, they resemble the old-fashioned sticks of confectionary sold as 'rock' at seaside resorts, in which the name of the resort is present throughout the length of the candy. The internal tubular structures are termed canals and their origin is explained by the presence of finger-like projections from the wall of the alimentary canal, immediately following expulsion from the stomach. These caused grooves to be excavated into the cylindrical faecal pellet as it moved through the gut prior to being evacuated. The projections are considered to represent organs that efficiently extracted nutriments and water from the pellet. During the passage of the pellet through the digestive system, the presence of internal projections decreased, allowing the grooves to be closed and form internal canals within the pellet. These microcoprolites are most commonly assigned to the genus *Favreina*.

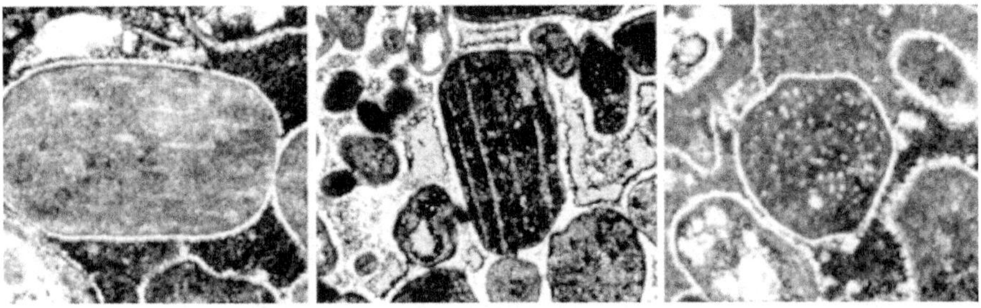

Left and middle: longitudinal sections of *Favreina salevensis* showing internal tubes. Right: tranverse section, showing tube patterns of taxonomic value.

Associated hypersaline-tolerant microfossils indicate that the decapod shrimp creators of *Favreina* lived in shallow marine conditions where salinity was often higher than normal and possibly food availability was lower than normal. Under such conditions, the development of internal gut projections would have proved vital to extract the maximum goodness from the food plug.

### *Prethecoprolithus*

Another distinctive microcoprolith found in limestones is *Prethecoprolithus*, characterized by cross sections that resemble a pair of spectacles or simple solid elongated rectangles. These are faecal ribbons with more elongate cross sections and an absence of internal canals. They were produced by bivalves, and similar forms are produced today by the common mussel *Mytilus*. As with *Favreina*, the microcoprolite shape represents a mould of an internal gut adaptation to efficiently extract nutriments and water from the food mass. This process is not unlike that used by playdough-extruding toys to create ribbons of varying shape. It does this by increasing the internal surface area of the gut and the decreased water content leads to increased rigidity. A distinct advantage of the rigid microcoprolite ribbon is the avoidance of pollution of the water, a vital adaptation in cases where the inhalent water current passes close to the anus.

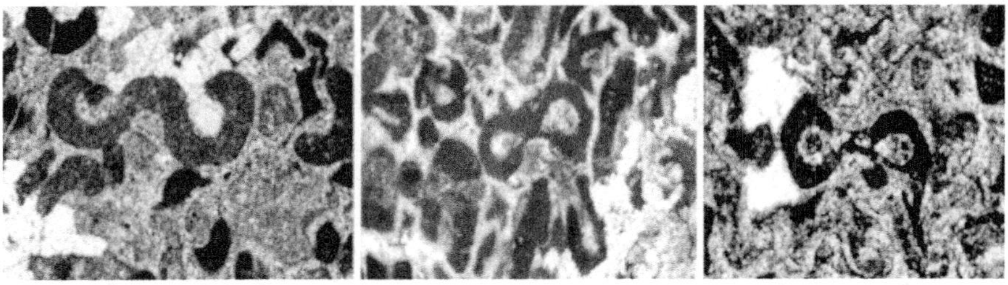

Transverse sections of ***Prethecoprolithus centripetalus*** reflecting the double loop alimentary canal.

*Prethecoprolithus* is typically found in limestones containing cyanobacterial microgranules — often in stromatolitic layers — and simple cyprid ostracods, and often close to evaporite beds. A bacterial grazing mode of life within a stressed, marine environment with elevated salinity and temperature is suggested.

The presence of such microcoprolites in the absence of typical normal marine microfossils is a useful indicator of possibly subnormal, shallow marine conditions. Their lateral distribution in any regional study of reservoir limestones can help in producing a palaeoenvironmental map that would assist in finding the best-quality hydrocarbon reservoir. Their distribution within a vertical rock succession can often reveal otherwise hidden sea-level fluctuations and associated depositional cyclicity, and also contribute to improve understanding of the three-dimensional distribution of reservoir fabrics. They can even be age indicators in limestones where conventional microfossils are absent.

# Molecular fossils

**Iain Scotchman**

'Not another fossil f***er!' was the greeting I got back in 1986 from the Amoco company man. I'd just arrived on the Sedco-703 drilling rig, to begin my first stint offshore. But I'm not and never have been a palaeontologist — I'm an organic geochemist. So, what was I doing there?

Well, micropalaeontologists might think that they are looking at the smallest fossils with their bugs and spores, but they are not. This distinction goes to the biomarkers of the organic geochemist. These are complex molecular fossils, and they tell us about some of the most important changes that took place in Earth's biosphere through time, in addition to having myriad uses in the petroleum industry and elsewhere.

Biomarkers are derived from biochemicals, particularly lipids, generally from bacteria, algae, and higher plants. They occur as preserved organic matter in sediments, and in the petroleum generated from them. High-tech instruments are required for biomarker analysis. We use techniques such as linked gas chromatography–mass spectrometry to study low concentrations of such compounds with complex molecular structures of more than 20 carbon atoms. Growth in applications of biomarker technology to petroleum geology, environmental studies, and archaeology has been rapid, extending even to more esoteric fields such as forensics, particularly related to narcotics (Peters et al. 2005).

Petroleum systems analysis has spurred development of biomarker techniques: much geological information on the source rock of an oil can be gleaned from a study of its biomarkers, ranging from identification of the organic matter to its depositional environment and geological age. By comparing with present-day systems, we can infer depositional environment (non-marine lacustrine to fluvial or marine) and water stratification (salinity, anoxia). Information on the geological age of the source rock can also be determined from age-specific biomarkers such as oleanane — related to flowering plants and indicating a late Cretaceous or younger age — and other biomarkers related to diatoms, which can further refine ages of Tertiary oil-source rocks. Even the beginnings of early life forms are indicated by unique biomarkers from Vendian–Cambrian reservoired oils potentially related to plankton and algae of Ediacaran age (Ivanova & Kashirtsev 2010).

*[Biomarkers] are complex molecular fossils, and they tell us about some of the most important changes that took place in Earth's biosphere through time.*

Thermal maturity of the source rock at the time of hydrocarbon generation, a critical factor in petroleum system analysis, can be determined from sterane and hopane biomarker ratios, although these are confined to levels of maturity in the so-called oil window. However the recent discovery of diamondoids (the 'bucky-balls' beloved by physicists) in high-maturity hydrocarbons such as condensates has elevated maturity studies into the realm of high temperature oil-to-gas cracking reactions. Even in biodegraded oils, biomarkers can survive to provide source and maturity information, as well as providing a useful measure on the quality of the oil based on the relative losses of various biomarker compounds as biodegradation reactions progress.

Outside of petroleum geology, oil-related biomarkers are used in environmental and even archaeological studies. For example, oil spills can be characterized and the source, and possibly the culprits, identified. In archaeology, bitumens and other hydrocarbons have had a multitude of uses since ancient times, generally for waterproofing but also for mummification. Biomarker analysis can determine the source of these bitumens, potentially constraining where early civilizations developed these techniques and their migration over time. Biomarkers of ancient resins and waxes, and even early narcotics, can give valuable information on early cultures and their development.

So the world's smallest fossils punch well above their weight, having a multitude of uses and applications, not only to the oil industry. Further rapid growth in their analysis and utility seems certain.

## References

Ivanova, I, and V Kashirtsev (2010). Distribution of monoalkylbenzenes $C_{12}H_{18}$–$C_{27}H_{48}$ in Vendian–Cambrian oils of the Siberian Platform. *Russian Geology and Geophysics* **51**, 1199–1203.

Peters, K, C Walters, and J Moldowan (editors) (2005). *The Biomarker Guide, Volumes 1 & 2*, Cambridge University Press, Cambridge, UK.

# Museum or mantlepiece: a moral dilemma

**Tim Palmer**

I have only ever found one dinosaur in my life. It just so happened that it was on the very day that I started my doctorate, on my first morning in the field in the Oxfordshire Great Oolite Group. There were two huge vertebrae, one very fragile and the other robust enough to be extracted in a plaster jacket by Philip Powell, and taken to the University Museum. I don't know whether it is still regarded as a *Cetiosaurus*, but that is how we thought of it at the time. Certainly it was a large sauropod, and it even got a paragraph in the *Oxford Times*.

The bed was a fine clayey quartz silt, full of exquisitely preserved aragonite molluscs, charophytes, brackish and freshwater ostracods, and identifiable plant material of 'upland type' (designated so by Professor Tom Harris at Reading University, the expert on the Middle Jurassic plants of the Yorkshire Moors and coast). The underlying clay was pierced by rootlets and was drawn up into the overlying bed in large flame-like structures that I later realized probably represented dinoturbation. Clearly we were dealing with a brackish swamp close to the shore of the Bathonian London Landmass (as it used to be called) where huge beasts roamed and trampled. The bed even got christened the 'Monster Bed' by Hugh Torrens.

Embedded in the silt close to the vertebrae were two large pebbles of dark grey-brown quartzite, several centimetres across and at least five orders of magnitude larger than any other sediment particle in the bed. They are both well-rounded to very well-rounded and it seems to me that they are very unlikely to be anything other than gastroliths. Like many other people who have been lucky enough to spend their working lives doing palaeontology, the first appeal of fossils to me was romantic and aesthetic rather than coldly scientific, and these two gastroliths have become a favourite possession. They are on my desk as I write this. I enjoy looking at them, and I particularly like to hold them in my hand as if they are worry beads. They click and clatter and squeak together, like some atheist's rosary (any characterization of the microscopic wear patterns on dinosaur gastroliths will not come from these specimens). My weird fondness for these superficially unremarkable objects lies solely in their extraordinary context and my memory of their discovery, and therein lies the moral dilemma. They ought to be in a museum, for I suspect that sauropod gastroliths

are not that common in northwest Europe. But sure as hell I do not want to put them there, where no one will ever think that they are romantic objects again, and where, at worst, no-one will ever look at them. It is even possible that some 'expert' may decide that they cannot be what I think they are (there are no drawers in natural history museums labelled 'dinosaur gastroliths'), and they will find their way into a skip.

It's not as if I am unwilling to share their curiosity value. My children were charmed by the allure of their story, even though, when young, they were disinclined to show too much interest in other aspects of geology. In first year tutorials, these two quartzite pebbles were an excellent basis for a discussion of grain shapes and sizes, and the importance of context. They even induced in some students, as they had in me, a feeling of thrill and awe — much more important than mere knowledge. That can always come later.

So I do not know what I am going to do with them. For the moment they can go back into the bowl on the mantlepiece, together with the huge *Torquirhynchia inconstans* from Ringstead Bay in Dorset and the enrolled trilobite from Ohio. In the longer term, the moral dilemma can simmer, in the hope that one possible course of action eventually prevails over the other. But it may very well not.

## References

Palmer, T (1973). Field Meeting in the Great Oolite of Oxfordshire. *Proceedings of the Geologists' Association* 1973, 53–64.

# Ordovician on the roof of the world

## David Harper

Mount Everest is the most iconic natural structure on Earth. Rising majestically to a height of nearly 8850 metres above sea level — an ascent is a true pinnacle of human achievement. Its unique climate, fauna, and flora have attracted the term, 'the planet's third pole'. The region is also home to many distinctive and endemic cultures, adapted to conditions on the high Himalayan plateau.

Although the initial reports by some of the early Everest climbing expeditions in the 1930s suggested that the limestones on the summit of Everest were Permian in age, intensive study by geologists from the Chinese Academy of Sciences in the mid 1970s established a new stratigraphic framework for the northern slope of the mountain, correlating the limestones on the summit with the Lower and Middle Ordovician elsewhere in the world, such as South China. The entire upper part of Everest (see figure), above the distinctive Yellow Band, consists of these Ordovician carbonates — over 200 m in thickness and assigned to the Chiatsun Group. We know little of the biota in these rocks, but fortunately they crop out at lower, more accessible altitudes in the Tibetan county of Nyalam, not far from the border with Nepal.

My own journey to the Everest Range was one of serendipitous coincidence: first, a PhD on the taxonomy of the Late Ordovician brachiopods of the Girvan District comprising some 500 pages of systematic descriptions, a route no longer fashionable for a career in palaeontological research; second, many years of collaboration with colleagues in the Chinese Academy of Sciences, Nanjing; and finally, a mid-career move to a professorship in the University of Copenhagen coupled subsequently with many field seasons working in Greenland. My research in South China had focused on a key event in Earth's history, the Great Ordovician Biodiversification Event (GOBE), when the planet's biotic diversity escalated over a relatively short interval of time.

A joint collaborative program between China and Denmark (largely sponsored by the Danish Innovation Centre in Shanghai) offered the opportunity to develop a new multidisciplinary project involving botany, geology, and zoology between our respective colleagues together with staff and students in the Tibetan University in Lhassa. During a couple of weeks of driving through some of the most hair-raising passes, reaching altitudes of 5000 m, absorbing some of

Left: View of the North Face of Everest showing the main structural and stratigraphic units (courtesy of Rick Law) — the shelly faunas are from the limestones on the summit. Right: Selection of the diverse brachiopod fauna from the Chiatsun Group.

the most amazing scenery on the planet, we sampled a number of key localities, the most productive occurring along the Friendship Highway.

Our new investigations, including detailed study of our fossil collections, have revealed a moderately diverse shelly fauna dominated by brachiopods, commonly in shell concentrations and deposited in nearshore carbonate environments. The new fossil data confirmed some of the previous Middle Ordovician correlations for this unit, but it also initiated monographic description of the brachiopods and interpretation of their palaeoecology, permitting a more precise correlation of this part of the Chiatsun Group. We also sampled some stratigraphically higher units within the group, resulting in a greater age range for the group than previously reported. Biogeographical analysis of the fauna suggests it may be related to the low-latitude Toquima–Table Head province that extended around the margins of the ancient continent of Laurentia (North America). This implies that a similar tropical belt was developed on the Tibetan margin of the continent of Gondwana during the Mid Ordovician.

What did our new geological investigations really achieve? We confirmed that the summit of Everest was indeed part of an ancient sea bed some 450 million years ago, that was thrust into its present position during the more recent collision of India with Eurasia. Our new data, however, are also constraining new models for the diversity of the marine faunas, their habitats and environmental conditions, together with the precise location of this part of Tibet on the globe during the Ordovician. The GOBE was a global event that also affected the Tibet margin of Gondwana in the shallow-water tropical conditions that prevailed at the time. But what a great opportunity to work on the roof of the world surrounded by an ancient Ordovician seafloor.

# Ostracods are amazing

**Vincent Perrier**

Ostracods are crustacean arthropods, typically around 1 mm long, though some planktonic 'giants' reach 30 mm. Their body is enclosed in a bivalve carapace made of calcium carbonate, into which the animal can retreat. They have colonized aquatic environments, from the ocean abyssal plains to damp leaf litter, and can tolerate a range of temperature, salinity, and other chemical stresses that would cook, pickle, or dissolve most other organisms. Ostracods are key components of aquatic ecosystems, as primary and secondary consumers.

The ostracod carapace is a success of biological engineering; light and streamlined in agile planktonic species, strong and ornamented in their benthic cousins to endure the pressures of living in the deep sea. This carapace is easily fossilized in the sediments — explaining why ostracods are the most numerically abundant arthropods in the fossil record. Known from 65 000 living and fossil species, ostracods have been present on Earth from the earliest Ordovician onwards, surviving nearly 500 million years of Earth history.

The ostracod carapace is very important for controlling the ambient aquatic environment within the carapace. It also protects the animal's eggs and juveniles providing a considerable advantage. This capacity for parenting is ancient and can be recognized from distinctive carapace features (e.g. the brooding pouch) in the Early Palaeozoic. Sometimes the fossilized eggs are miraculously preserved inside the female carapace, as in the Silurian ostracod *Nymphatelina gravida* from the Herefordshire Lagerstätte (Siveter et al. 2007). Another exceptional fossil from Herefordshire, *Colymbosathon ecplecticos*, preserves its penis, thus being the oldest known male reproductive organ in the fossil record (Siveter et al. 2003). Ostracods also have one of the longest sperm cells in the animal kingdom: it can be up to 10 times longer than the male body. This giant sperm reproductive strategy is known at least from the Miocene and possibly from the Cretaceous (Matzke-Karasz et al. 2014). These rare exceptional fossils show morphologies very similar to living forms, demonstrating the stability of the group for hundreds of millions of years.

The incredible abundance of fossils reveals their capacity for colonizing new environments. The ostracods that colonized the water column have particular morphologies that support their advanced swimming capacity. The Herefordshire fossils show that these anatomical traits were developed early in the group

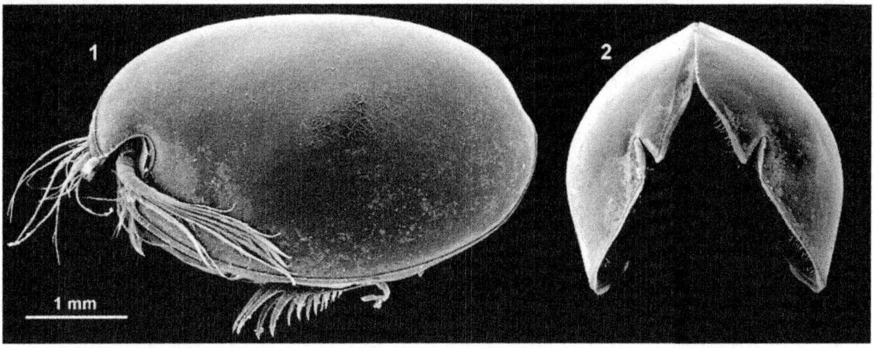


Two SEM views of the recent nektobenthic ostracod *Vargula hilgendorfii*.

and, importantly, prior to fossil evidence of ostracods colonizing the plankton. Equipped with strong swimming appendages, powerful vision, and efficient respiration, these precursors to the first planktonic ostracods may just have needed a small push off the seabed to invade the plankton. This push may have happened 427 million years ago, after a major extinction event that especially affected the global plankton (graptolites, conodonts, and chitinozoans), and is known colloquially as the 'Big Crisis'. It was after this that the ostracods entered the water column, possibly taking advantage of a new food source or low predation pressures post-extinction, or were pushed off the seabed by environmental changes associated with the Big Crisis (Perrier et al. 2015).

The benthic ostracods may have colonized freshwater habitats at least seven times in the geological record. They were the vanguard of colonizing non-marine aquatic environments, taking advantage of an abundant new food source: land plants. Early attempts to colonize brackish waters occurred repeatedly during the Silurian and Devonian. However, from the Late Devonian onwards there were marked increases in the frequency of these radiations, and by the Late Carboniferous ostracods were firmly established in freshwater habitats (Bennett et al. 2012). The onset of major glaciation in the Carboniferous may have triggered the ostracod colonization of the land, for example by sea-level change isolating water bodies that freshened over time.

## References

Bennett, C, D Siveter, S Davies, M Williams, I Wilkinson, M Browne, and C Miller (2012). Ostracods from freshwater and brackish environments of the Mississippian of the Midland Valley of Scotland: the early colonisation of terrestrial water bodies. *Geological Magazine* **149**, 366–396.

Matzke-Karasz, R, J Neil, R Smith, R Symonová, L Morkovsky, M Archer, S Hand, P Cloetens, and P Tafforeau (2014). Subcellular preservation in giant ostracod sperm from an early Miocene cave deposit in Australia. *Proceedings of the Royal Society, Series B* **281**, 20140394.

Perrier, V, M Williams, and D Siveter (2015). The fossil record and palaeoenvironmental significance of marine arthropod zooplankton. *Earth-Science Reviews* **146**, 146–162.

Siveter, D, M Sutton, D Briggs, and D Siveter (2003). An Ostracod crustacean with soft-parts from the Lower Silurian. *Science* **302**, 1749–51.

Siveter, D, D Siveter, D Briggs, and M Sutton (2007). Brood care in a Silurian ostracod. *Proceedings of the Royal Society, Series B* **274**, 465–69.

# Palaeontology and the public consciousness

## Rachel Preece

Palaeontology — such an impossible tongue twister of a word that almost every five year old appears fully conversant with. Is it possible to find a child's book-shelf without at least one dinosaur book? Indeed, palaeontological imagery abounds in the public consciousness. One does not to have to look far, on the high street or in the media, to spot fossil symbolism.

While not strictly a fossil, there is no getting away from the ubiquitous Shell logo. What started out as a small London antiques and oriental seashell business became the trademark for kerosene shipped to the Far East by Marcus Samuel and Company in the 1890s. The first logo of the re-branded Shell Transport and Trading Company was a mussel shell. In 1904 this was changed to a scallop (*Pecten*) and subsequently appeared on every tanker carrying kerosene to the Far East; in fact each ship was named after a different seashell.

Perhaps less well recognized, but up there in the eccentricity stakes, is the Sinclair Oil Corporation of Utah, USA. Founded by Harry F Sinclair in 1916 the corporation's logo features the silhouette of a large green dinosaur (an *Apatosaurus*, controversially also known as a *Brontosaurus*). At the Chicago World's Fair of 1933–34, Sinclair sponsored a dinosaur exhibit, under the misconstrued correlation between the formation of petroleum deposits and the time of dinosaurs. The exhibit included a two-ton animated model of a *Brontosaurus* and proved so popular it inspired a promotional line of wiggling rubber *Brontosaurus* at Sinclair gas stations and the eventual inclusion of the *Brontosaur* logo. The logo is parodied in the *Toy Story* movies as being the logo for the 'DinoCo' gas station chain.

Nautiloids and ammonites are rich pickings for graphic designers in search of an easy-on-the-eye corporate identity. From corporate training providers to mineral exploration companies, film production, and even the North Korean Institute for Security and Sustainability, the nautiloid seems to embody corporate visions and cultures.

Sports teams love a good assertive branding. The Toronto Raptors basketball-dribbling dinosaur logo is now replaced with a sleek silver basketball, ripped by dinosaur claw marks. Hamburger SV is a sports club and home to one of

Germany's oldest and most famous football teams with the unique distinction of having played continuously in the top tier of the country's football league since the end of World War I . They are known affectionately as *der Dinosaurier* (the Dinosaur). The University of Calgary athletic teams are known as 'The Dinos', and the Colorado Rockies baseball team has a blue *Triceratops* mascot, whereas both the Nashville Predators and the Buffalo Sabres ice-hockey teams favour a *Smilodon* (sabre-toothed cat).

No reflection on fossils and unconscious awareness can be complete without mention of Lyme Regis in Dorset, England, and its spectacular ammonite street lamps. This aesthetic nod to the area's geological heritage was completed after intensive public consultation.

Kemmerer, Wyoming, USA is touted as 'An aquarium set in stone' due to its proximity to Fossil Butte National Monument. The town is home to the first JC Penny store but of greater note it also celebrates its natural heritage through sidewalk plaques commemorating the fossils found in the region.

While London's Natural History Museum may claim to be the birthplace of fossil symbolism, the Black Country Geological Society of Dudley, England, should not be overlooked. Dudley features the iconic 'Dudley Bug,' *Calymene blumenbachii*. This trilobite is special to the Black Country and, historically collected from the local limestone mines, it was featured in Dudley's former coat of arms. However, even this achievement in palaeontological promotion is surpassed by the USA's state fossils — one for almost every state.

Why? Why are these images so readily available and fully adopted within everyday life? Is it the intrigue of a life long gone and the perpetuating mystery? Is it a desire to share the excitement of a rare and exclusive find? Is it an appreciation of the beauty in a natural form? Or perhaps a reflection of the uncomfortableness of not fully knowing? That irksome itch of not having the answer. Do we all share a sense of 'trying to make sense' of it all?

Do we dilute the magnitude of these imponderables within our imaginations through sharing the images? Are we avoiding embracing the fragility and uncertainty of our own world and subsequently, with the unintended — but delightful — consequence, of consciously and subconsciously programing kids from an early age to recognize fossil forms and celebrate them… whimsically.

# Phylogenomic tools for molecular palaeobiology

**Gert Wörheide**

Molecular palaeobiology is a relatively new field within palaeontology. It uses molecular biology methods to address earth science questions for a more complete understanding of biosphere evolution through deep time. It uses genomic information from extant organisms (Wörheide et al. 2016), but does not use molecules preserved in the fossil record, such as lipid biomarkers or ancient DNA.

The field greatly benefits from methodological revolutions in molecular biology, most recently massive parallel sequencing (aka next generation sequencing, NGS), including genomics and transcriptomics. The cost of high-throughput sequencing is steadily dropping and new NGS sequencers, such as Illumina's MiniSeq or Oxford Nanopore's MinION, are smaller and more affordable than ever. Consequently, such methods are now increasingly applied in palaeobiological and geobiological research.

A hot topic in molecular palaeobiology research concerns the emergence and early evolution of animals. It is still not fully understood when animals first evolved, because molecular clock studies suggest much earlier dates than the fossil record documents. The branching order of the main animal lineages is also intensively discussed. However, resolving the correct branching order of the non-bilaterian lineages (Porifera, Placozoa, Ctenophora, Cnidaria) and their relationships to the Bilateria, which include all other animals, is important not only to reconstruct the last common ancestor of all animals, but also to elucidate the early evolution of key animal traits, such as the nervous system and muscles.

The early fossil records of some of the key lineages, especially the Ctenophora and Placozoa, are scarce or non-existent so the geological record is of little use to resolve phylogenetic relationships or trait evolution. However, the classical model of animal evolution, based on cladistic morphological character analysis of extant taxa (Ax 1996), poses that nerve- and muscle-lacking sponges (Porifera) were the earliest branching animal lineage, implying a single origin of neurons and muscles in the last common ancestor of Ctenophora, Cnidaria, and Bilateria.

Another long-standing question is whether sponges (Porifera) are monophyletic (as suggested by cladistic analysis of morphological characteristics) or instead a paraphyletic assemblage that shares a grade of construction rather than common ancestry, as suggested by numerous early gene analyses (reviewed in Wörheide et al. 2012). This hypothesis has gained momentum especially among palaeontologists who could easily agree with a stepwise emergence of sponge lineages ('classes') because indeed they appear subsequently in the fossil record. However, paraphyletic sponges would have profound implications for the understanding of our own species heritage. If so, we all would be evolved from sponges because the last common ancestor of sponges and the rest of all animals, including us, would have been a filter-feeding organism with an aquiferous system!

Fortunately, in the last decade, progress in massive parallel sequencing has nurtured the new field of 'phylogenomics', which uses large genome-scale multi-gene matrices to reconstruct deep phylogenetic relationships. While phylogenomics has contributed significantly to stabilizing the bilaterian tree of life, some studies provided new controversial hypotheses about the early evolution and diversification of non-bilaterian animals (reviewed in Dohrmann & Wörheide 2013). Among those, the 'Ctenophora-sister' hypothesis is the most fiercely debated one, proposing that the comb jellies (Ctenophora) are the earliest branching animals and not the sponges. If so, perceptions about a stepwise evolution of complexity must be rethought, because ctenophores are morphologically complex, mainly pelagic predators with muscles and nerves. Nonetheless, other studies recovered a more traditional arrangement at the root of the animal tree of life, with sponges as the first offshoot (the 'Porifera-sister' hypothesis), and a recent study suggested that Ctenophora-sister is likely a phylogenetic tree reconstruction artefact (Pisani et al. 2015) although this still is a point of contention (Telford et al. 2016). All recent studies agree, however, on the monophyly of sponges, and that is at least something.

Despite the heated discussion over which lineage is the earliest branching one in the animal tree, it has initiated a critical re-evaluation of classical evolutionary scenarios, along with much novel and valuable research on neglected non-bilaterian groups, such as sponges, ctenophores, and placozoans. If only the fossil record would be more phylogenetically informative with respect to those lineages.

## References and further reading

Ax, P (1996). *Multicellular Animals: A new Approach to the Phylogenetic Order in Nature*. Springer, Berlin.

Dohrmann, M and G Wörheide (2013). Novel scenarios of early animal evolution — is it time to rewrite textbooks? *Integrative and comparative biology* **53** (3), 503–511.

Pisani, D et al. (2015). Genomic data do not support comb jellies as the sister group to all other animals. *Proceedings of the National Academy of Sciences of the United States of America* **112** (50), 15402–15407.

Telford, M, L Moroz, and K Halanych (2016). Evolution: A sisterly dispute. *Nature* **529** (7586), 286–287.

Wörheide G, M Dohrmann, D Erpenbeck, C Larroux, M Maldonado, O Voigt, C Borchiellini, and D Lavrov (2012). Deep phylogeny and evolution of sponges (Phylum Porifera). *Advances in Marine Biology* **61**, 1–78.

Wörheide, G, M Dohrmann, and Q Yang (2016). Molecular palaeobiology — Progress and perspectives. *Palaeoworld* **25** (2), 138–148.

# Remove the rock to reveal the past

**Gunn Mangerud & Niall Paterson**

When strolling along a river in a forest or walking along the seaside after a storm, you might notice remains from plants and trees floating on the water. As well as the twigs and leaves you can easily see, there are spores and pollen that you probably cannot. Although most of this organic material will be lost through the processes of microbial decay and oxidation, some of it will be carried out into a lake or the sea where it will eventually sink to the bottom and be buried by sedimentary particles. Some of these plant remains will, after being embedded, become part of the sedimentary rock.

Typically, the accumulation of organic matter is restricted to aquatic environments, because in subaerial settings plant particles are rapidly destroyed by microbial action. Once enclosed in the sedimentary rock, organic debris might be preserved over geological timespans, thus forming an archive to the terrestrial vegetation that once grew in the region. Such information is incredibly valuable in reconstructing palaeoenvironments, and also in assessing the hydrocarbon potential of the host rock through a technique often referred to as biofacies or palynofacies analysis.

In order to study the organic remains of a sedimentary rock the organic debris is first isolated by removing all the rock material through various processing techniques using hydrochloric and hydrofluoric acids, leaving only the acid-resistant organic particles behind. This material includes pollen and spores, parts of leaves, wood fragments, fungal hyphae, lichens, algae (both marine and fresh water), and particulate organic components that appear structureless at the scale of light microscopy. We call this last class amorphous organic matter.

The quantity and quality of the organic matter that eventually becomes preserved in the rock are the result of the total biomass productivity, both onshore around the basin, and offshore. In addition, degradation, sedimentary transport processes, and relative sea level all influence the ultimate composition of the preserved organic matter assemblage.

By performing a visual analysis of the organic matter composition present in a rock, a palynologist can provide an estimate of the relative distance from the palaeoshoreline. Typically, large and heavy organic particles have a tendency to

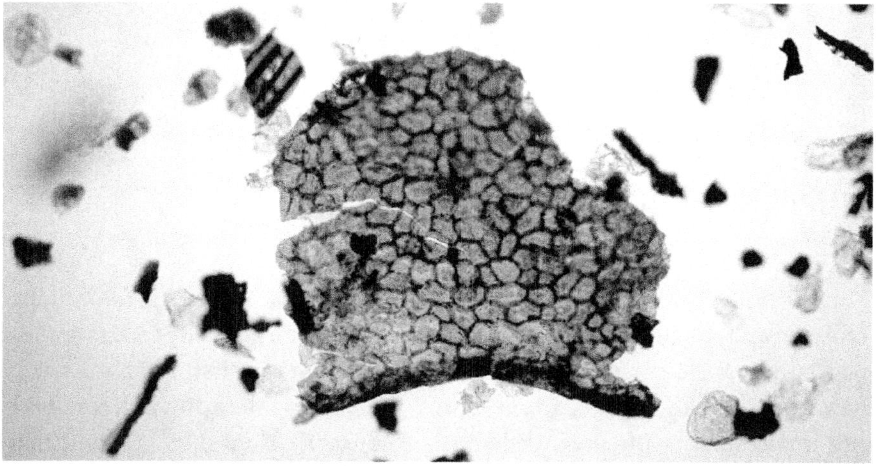

The image shows 220 million-year-old organic remains, including a cuticle fragment from the Snadd Formation, offshore Kong Karls Land, Barents Sea.

settle closer to shore. Natural sorting mechanisms therefore result in variation in the ratio of dense (e.g. fern spores) to lightweight (e.g. bisaccate pollen) organic particles along an onshore–offshore gradient. An increase in the abundance of pollen relative to spores indicates a change to a more offshore palaeoenvironmental setting. This change will often be accompanied by a corresponding increase in the relative abundance of marine palynomorphs (e.g. acritarchs and dinoflagellate cysts), as well as a change in the organic composition, such as an increase in amorphous organic matter.

There are economic implications of palynology. Since terrestrially sourced organic material is associated primarily with a gas-generating rock, and amorphous and algal-derived organic matter with a typically oil-prone rock, visual organic matter analysis can also provide a quick means of assessing the petroleum potential and maturity of a source rock.

The composition of the pollen and spores contained within an organic matter assemblage can also be used to reconstruct the composition of the palaeoflora within a given region. By grouping palynomorphs according to their presumed botanical affinity, based upon records of *in situ* occurrences within parent plant taxa, they act as proxies for the onshore vegetation. Because plants adapt to different habitats we can infer the preferences of the parent plants and give valuable evidence for the environment they grew in. For example, the spore *Leschikisporis aduncus* was produced by ferns, which are typical of wet lowlands. In conjunction with geological and geophysical information this can be a powerful tool to reconstruct the palaeoenvironment at various times through the geological record.

# Scottish Carboniferous Fossil-Lagerstätten

**Euan Clarkson**

Lobsters and crabs are familiar to us all. Yet all these diverse kinds of higher crustaceans (eumalacostracans) had common ancestors, and we can trace these back as far as the Late Devonian. They underwent their first, dramatic adaptive radiation in the Early Carboniferous. But because they do not have purely calcareous shells — their exoskeletons are constructed mostly of organic matter — they needed special conditions for preservation, and such circumstances only prevailed occasionally.

Such conditions were provided by the remarkable palaeogeography of the Midland Valley of Scotland, This is a graben bounded by the Highland Boundary Fault to the northwest and the parallel Southern Upland Fault to the southeast, and great thicknesses of sediment accumulated therein from Silurian to Carboniferous times. In the eastern part of the country fresh and brackish lakes prevailed, sometimes subject to marine influence. The two localized areas of sedimentary rock that concern us here are of Lower Carboniferous age; they are the Gullane and Granton 'shrimp beds', exposed along the southern shore of the great estuary of the Firth of Forth. These Fossil-Lagerstätten are thin, layered beds of dolomitic sediment within shale sequences, and they contain exquisitely well-preserved eumalacostracan crustaceans of distinctly modern appearance.

The Gullane shrimp bed, some 10 cm thick, is brown in colour. It has one species only, *Tealliocaris woodwardi*, extremely similar in appearance to modern shrimps. T H Huxley remarked on the similarity, with evident surprise, as long ago as 1847, when he described the first-known Carboniferous shrimp. Adult *Tealliocaris* specimens can be up to 4 cm long, but juvenile individuals are not uncommon. The shrimps may be preserved in dorsal, ventral, or lateral orientation, which allows such anatomical features as the antennae, legs, eyes, and the tail fan to be clearly seen. These features are preserved in fluorapatite, originally deposited as a stable bacterial film which rapidly spread over a dead shrimp, retaining the structure. The Gullane shrimp bed was probably deposited in a thermally stratified freshwater lake at the margin of a much larger lake lying to the north.

The best way of studying the shrimps was to use the camera lucida microscope, which enables the observer's pencil to be seen superimposed upon the image of the fossil. This allows accurate drawings to be made, from which reconstructions can be prepared. In some layers the exoskeleton is preserved almost unchanged, revealing a wealth of structural detail with the scanning electron microscope.

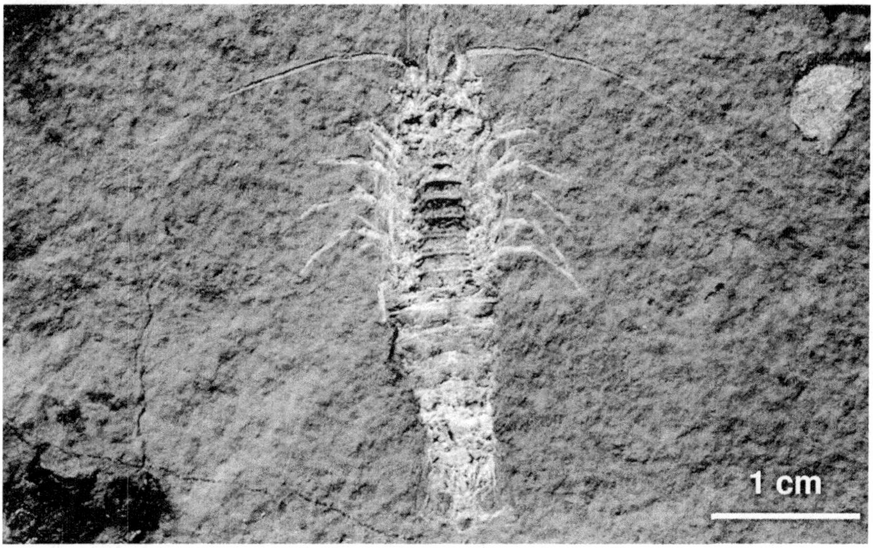


*Tealliocaris* ×4 from the Gullane shrimp bed.

The Granton shrimp bed is lithologically quite similar, but it is black. The fauna is much more diverse and occurs in several layers, probably mass mortality horizons. It is dominated by the small (1–2 cm), long and thin shrimp *Waterstonella*, which occurs in hundreds on the bedding planes. More rarely seen is the five-times larger predator *Palaemysis*, otherwise known as *Anthracophausia*. Here the second antennae are modified as a catching apparatus, like jackknives with inward facing spines. Living mantis shrimps have similar equipment, though it is the mouthparts that are modified. There are also some *Tealliocaris* specimens, other kinds of shrimps, large ostracodes, the first-described conodont animals, the earliest-known tomopterid worms, and occasional large nautiloids. It seems to have been an algal mud-flat environment at the margin of a fresh- or brackish-water lake in which most of the shrimps lived. But it was not far from the sea, and was subject to occasional storm surges which brought in the marine organisms, and killed the shrimps.

So, we have these two Fossil-Lagerstätten which give us a fascinating picture of eumalacostracan life some 350 million years ago. The shrimps are extraordinarily similar to those living today, epitomizing good design. But not in all respects: none of the Carboniferous shrimps have claws!

# Symmetry in early echinoderms

**Imran A Rahman**

Starfish are among the most familiar animals encountered on the seashore today. Together with other well-known forms such as sea urchins, sea cucumbers, and brittle stars, they belong to a major group of marine animals called the echinoderms (meaning 'spiny skin'). The group is characterized by five-fold symmetry, which means they can be split up into five roughly equal parts. In contrast, most other animals are characterized by bilateral symmetry, meaning they can be divided into mirror-image halves. Because five-fold symmetry is unique to echinoderms, uncovering how this type of symmetry evolved has proven especially challenging for scientists.

Palaeontologists can turn to the fossil record to get a better understanding of the early evolution of echinoderms. The oldest fossil echinoderms come from the Cambrian period, over 500 million years ago. These include extinct forms that were bilaterally symmetrical and asymmetrical, as well as taxa that have three-fold symmetry or five-fold symmetry. We know these fossils are echinoderms because they all possess a hard, mineralized skeleton composed of calcium carbonate; however, they are otherwise rather different from one another, and so it can be difficult to work out their evolutionary relationships.

Recent research has produced a new tree depicting the relationships of different Cambrian echinoderms, which allows us to decipher the evolutionary history of symmetry in the group (see figure). The earliest echinoderms (e.g. *Ctenoimbricata*) were bilaterally symmetrical, like most living animals and the common ancestor shared by echinoderms and other animals. Subsequently, echinoderms evolved to be asymmetrical, as seen in *Asturicystis* and *Castericystis*. This was followed by a few forms that exhibited three-fold symmetry, like *Helicoplacus*. Lastly, echinoderms with five-fold symmetry evolved and diversified, giving rise to a range of different fossils including *Helicocystis*, *Camptostroma*, *Lepidocystis*, *Gogia*, and *Stromatocystites*. Future discoveries could overturn this tree, but it represents our best current hypothesis of how early fossil echinoderms are related.

The fossil record shows that the evolution of echinoderms with different types of symmetry occurred over a period of five to ten million years, which is — geologically speaking — very quick. This fits with what is documented for other

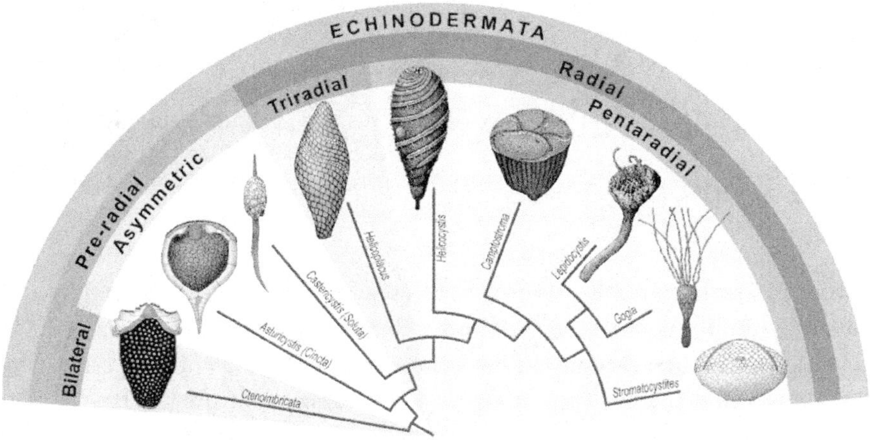


Phylogeny of select Cambrian echinoderms (Zamora & Rahman 2014).

animal groups, which appear abruptly in the Cambrian fossil record during the so-called Cambrian Explosion. Numerous hypotheses have been put forward to explain the Cambrian Explosion of echinoderms and other animals, ranging from increased oxygen levels to predator–prey arms races, but there is currently no consensus among scientists. For echinoderms, it seems likely that the ability to form a skeleton was important, and so it could be that changes to seawater chemistry played an important role in their origin and early evolution. Different types of symmetry were probably the product of developmental changes, i.e. the transformations that occurred during the transition from embryo to adult, but it is hard to be certain about this as nearly all the echinoderms preserved in the fossil record are adults.

Another open question is why echinoderms evolved five-fold symmetry, given that this type of symmetry is not seen in other animals. Some have suggested that five-fold symmetry aided feeding in early echinoderms, which are thought to have lived on the seafloor and fed on particles suspended in the water column. Testing this theory will be an important goal for echinoderm researchers over the coming years.

## References and acknowledgments

Zamora, S and I A Rahman (2014). Deciphering the early evolution of echinoderms with Cambrian fossils. *Palaeontology* November **57**, 1105–1119.

# The genius of Alcide d'Orbigny

**Mike Simmons**

Two themes run through my geological research — sequence stratigraphy and fossil foraminifera. The origins of both subjects trace back to one man Alcide d'Orbigny, who first recognized foraminifera as biological entities and understood their stratigraphic application. As a geologist, he introduced the concept of stages, the now standard means of chronostratigraphic subdivision. Moreover, he related these stages to global events — 'the expression of the boundaries which Nature has drawn with bold strokes across the whole globe' — immediately bringing to mind eustasy driven sequences. But this represents only a small part of d'Orbigny's scientific activity. Several years ahead of Darwin, he undertook a major eight-year expedition of South America focused on describing the plants, animals, and native people he encountered. On his return, he set about developing an atlas of all invertebrate fossils encountered in his native France, along with their associated geology. His scientific aims were never less than bold, and neither were his travels without adventure. He passed 15 days in a Uruguayan prison after local soldiers found his barometer suspicious.

Born 6 September 1802 in Couëron, France, he was only 11 years old when he began to examine foraminifera found in the local beach sands of the French Atlantic coast. These often beautifully shaped and complex microscopic shells became a subject of special interest to him, and in 1825 he used the term 'Foraminifera' to describe them, although initially mistakenly regarding them as tiny forms of cephalopod. His *Tableau méthodique de la classe des Céphalopodes* enumerated the various taxa he recognized in both modern and fossil forms, along with detailed descriptions of their morphology and mode of life. He also made physical models of many species of foraminifera, enlarged 40–200 times, so that their features could be more easily observed.

The efforts required to document his South American expedition (1826 to 1834) did not deter him from expanding his research into the geological record at home in Europe. In 1839 he described the foraminifera of the Cretaceous Chalk. Importantly, he surmised that during the Cretaceous the Paris Basin had been invaded by a warm sea lacking strong currents — a pioneering attempt at using microfossils for palaeoenvironmental purposes and to recognize transgression. Moreover, he noted, 'the comparative study of fossil foraminifera of all zones has proved to me a fact of geological importance: it is that each zone has its characteristic species, by which it may be recognized in whatever circumstances that can occur'. This was the first expression of the now routinely accepted concept that microfossils have strong biostratigraphic utility.

He devoted the later years of his life to his monumental *Cours élémentaire de paléontologie et de géologie stratigraphiques*. It is within these volumes that d'Orbigny outlined his views on stratigraphic organization, including stages (*étages*) as a fundamental concept — bodies of rock characterized by a particular set of fossils and typified by a particular location (a stratotype). He was building on the ideas of William Smith, but more particularly Baron Georges Cuvier, who explained the extinction of fossil organisms by means of episodic catastrophes. To d'Orbigny, each stage was separated by an upper and lower discordance representing an episode of natural upheaval. The preserved rocks representing the stages were, in his view, the result of transgression. Thus he viewed stages as transgressions separated by unconformities — a clear forerunner of sequence stratigraphic concepts.

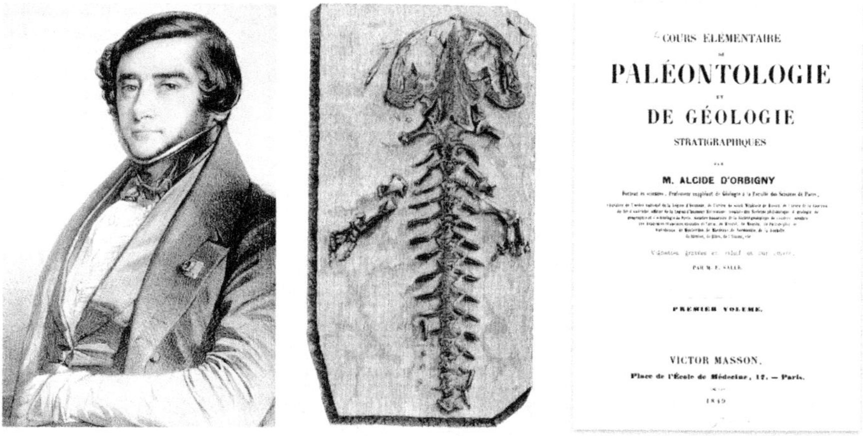

COURS ÉLÉMENTAIRE

PALÉONTOLOGIE

DE GÉOLOGIE

STRATIGRAPHIQUES

M. ALCIDE D'ORBIGNY

PAR M. F. SALLE.

PREMIER VOLUME.

VICTOR MASSON,

Place de l'École de Médecine, 17. — Paris.

1849

Alcide d'Orbigny from a lithograph by Lavallée; the Miocene amphibian *Andrias scheuchzeri*, as depicted in *Cours élémentaire*; and, right, the title page of *Cours élémentaire*.

In the years after his death in 1857 (at the age of 54) his views on catastrophism found little favour with geologists absorbing Lyell's uniformitarianism. But his ideas are now in tune with what we today call event stratigraphy, mass extinction, and sequence stratigraphy.

As palaeontologists and stratigraphers we owe a huge debt to d'Orbigny and can only marvel at the breadth of his scientific vision.

# The J sequence scheme

**Robin Dyer**

Exploration companies have drilled over 8000 wells in the UK sector of the northern North Sea. Happily, operators have been generous in allowing geological data and interpretations into the public domain through conferences and formal publications.

Publications have developed innovative ideas, encouraged scientific debate, and driven successful exploration and development of resources. Papers published by employees of BP in 1993 addressed the biostratigraphy, regional stratigraphy, and sequence stratigraphy of Jurassic sediments (Partington et al. 1993a; 1993b). From the viewpoint of an independent stratigrapher, these papers represent a significant release of data and ideas and rank as some of the most important publications of North Sea stratigraphy. This essay attempts to recognize and highlight the contribution they have made to unravelling the complex Jurassic stratigraphy of the region.

Jurassic sediments of the northern North Sea are important for two main reasons: the sandstones have yielded 75 percent of hydrocarbon production in the UK sector, and the mudstones are the main source rock for the basin.

BP established a large subsurface team in the late 1980s to undertake a major review of regional geology and prospectivity using new ideas related to sequence stratigraphy. An existing sequence stratigraphic scheme, adopted and published by Exxon, emphasized a depositional model driven by global falls in base level, resulting in basinward shifts in marginal facies which defined depositional sequence boundaries. This model works well along passive basin margins but the Jurassic rift basins of the northern North Sea posed a challenge and an alternative explanation of stratigraphic observations was required.

BP therefore adopted ideas developed in the Gulf of Mexico by Bill Galloway (1989a, 1989b). This scheme used the stratigraphic signals created by marine maximum flooding surfaces (MFSs) to mark genetic sequence boundaries and determine the stratigraphic filling history of basins. The Jurassic sediments of the northern North Sea were well suited to this approach, being deposited predominantly in marine environments in a rift basin. Differentiating the conflicting signals between regional eustatic sea-level variation and those resulting from local rifting was aided by extensive knowledge of the UK onshore Jurassic stratigraphy. Indeed, such was the calibration and correlation of offshore and onshore MFSs, the events were labelled after ammonite zonal index taxa, normally found only in onshore exposures and in continually cored boreholes.

Mark Partington and his colleagues, including John Underhill of Edinburgh University, defined eight tectono-stratigraphic sequences (J00 to J70). These were subdivided into 29 genetic sequences, all bounded by maximum flooding surfaces (J02 to J76) and defined primarily from petrophysical log data. Biostratigraphic calibration is absolutely critical to the correct definition of each MFS and the use of the genetic sequence model has elevated the importance of biostratigraphic data well beyond the standard use for age dating sediments.

I have tested and applied the J sequence scheme throughout the Viking Graben, the Moray Firth, and the Central Graben. It is a very powerful stratigraphic tool for correlation, reservoir mapping, and prediction. It provides a uniform stratigraphic nomenclature across the northern North Sea basins and has been adopted widely as an industry standard by many companies.

The J sequence scheme is not without flaws however. The calibration and definition of some boundaries are uncertain, as are the positions of some of the tectono-stratigraphic sequence boundaries. Nevertheless it has aged extremely well and remains one of the most significant geological contributions of the North Sea publication library. Every geoscientist working the Jurassic of the region should be well acquainted with the concepts and have the papers always within easy reach.



## References

Galloway, W (1989a). Genetic Stratigraphic Sequences in Basin Analysis I: Architecture and Genesis of Flooding-Surface Bounded Depositional Units. *AAPG Bulletin* **73**/2, 125-142.

Galloway, W (1989b). Genetic Stratigraphic Sequences in Basin Analysis II: Application to Northwest Gulf of Mexico Cenozoic Basin. *AAPG Bulletin* **73**/2, 143-154.

Partington, M, P Copestake, C Mitchener, and J Underhill (1993a). Biostratigraphic calibration of genetic stratigraphic sequences in the Jurassic-lowermost Cretaceous (Hettangian to Ryazanian) of the North Sea and adjacent areas. *Petroleum Geology of Northwest Europe: Proceedings of the 4th Conference* (Edited by J R Parker). The Geological Society, London. 371–386.

Partington, M, B Mitchener, N Milton, and A Fraser (1993b). Genetic sequence stratigraphy for the North Sea Late Jurassic and Early Cretaceous: distribution and prediction of Kimmeridgian-Late Ryazanian reservoirs in the North Sea and adjacent areas. *Petroleum Geology of Northwest Europe: Proceedings of the 4th Conference* (Edited by J R Parker). The Geological Society, London. 347–370.

# The last of their kind

## Nathan Van Vranken

Texas has one of the best records of deep time found in North America and a rich history to go with it. The rocks represent a variety of ages ranging from the Precambrian to the recent Quaternary and have been heavily sampled by palaeontologists and avid fossil collectors since the late 1800s. Several famous geologists and palaeontologists have passed through our state looking for the next big discovery. A few of them set up museums, excavations, or research institutions.

As a Texas-based palaeontologist, one of the subjects I enjoy most is studying the palaeoecology and biostratigraphy of the Western Interior Seaway, a large inland sea that covered the central part of North America from the Gulf of Mexico to Canada during the Cretaceous. The seaway tells a wonderful story about one of the more unique fossil representatives — the ichthyosaurs — and helps us understand their final days.

One of the fossils I enjoyed working on during graduate school was a tooth from west Texas, shown below. Not knowing what the tooth belonged to, it sparked a small research project. Over time I determined it was an ichthyosaur tooth and a west Texas first occurrence for a rare animal normally found in east Texas. The tooth is now housed at the Witte Museum in San Antonio.

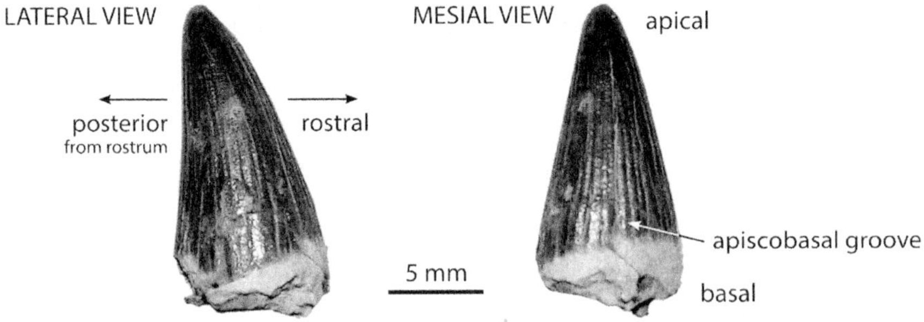


An *Opthalmosaurid* ichthyosaur tooth from west Texas. Described in Van Vranken (2017).

Ichthyosaurs are marine reptiles that evolved in the Triassic period and lasted up until the early Late Cretaceous. They achieved impressive evolutionary feats reaching some of the largest body sizes in the fossil record, such as *Shonisaurus*, and developing thunniform swimming motion

(Montani 1999; Nicholls & Manabe 2004). However, one genera that strikes me as fascinating is *Platypterygius*. *Platypterygius* was a 7 m long *Opthalmosaurid* ichthyosaur that once swam the oceans globally, but by the Late Cretaceous (93.5 Ma, Turonian) the genus becomes restricted to very few species within North America. Most of the remains for the genus found in North America are skull fragments, vertebrae, and the occasional tooth.

Texas comes into this equation because we have the rocks that preserve the latest occurrences of this ichthyosaur in abundance. Below, I ask a couple of questions that came up during my research which makes this topic even more rewarding.

1. **What drove ichthyosaurs extinct after being on the earth for so long?** Before the Cenomanian–Turonian boundary of the Cretaceous, ichthyosaurs had to deal with the changing environment and food sources, evolutionary conservatism, and competition from fish and other marine reptile groups such as early mosasauroids and plesiosaurs (Bakker 1993; Zammit 2012; Fischer et al. 2016). Evidence of these drivers can be seen in Texas strata. Ichthyosaurs were driven to extinction because they could not adapt any further.
2. **Were they as common as we think?** Ichthyosaur material may actually be misidentified in collections due to our lack of understanding of their morphology and occurrences. Myself and others seek to correct some of this by going into collections and refining the identification of the various bones. The Texan material relating to *Platypterygius* has been significantly overlooked and not actively studied until now.



## References and further reading


Adams, T and A Fiorillo (2011). *Platypterygius Huene*, 1922 (Ichthyosauria, Ophthalmosauridae) from the Late Cretaceous of Texas, USA. *Palaeontologia Electronica* **14**, 19A.

Bakker, R (1993). Plesiosaur extinction cycles — events that mark the beginning, middle and end of the Cretaceous. Evolution of the Western Interior Basin. *Geological Association of Canada, Special Paper* **39**, 641–664.

Fischer, V, N Bardet, B Roger, B Benson, M Arkhangelsky & M Friedman (2016). Extinction of fish-shaped marine reptiles associated with reduced evolutionary rates and global environmental volatility. *Nature Communications* **7**, doi:10.1038/ncomms10825.

McNulty, C and B Slaughter (1962). An ichthyosaurian centrum from the Albian of Texas. *Journal of Paleontology* **36**, 346–347.

Motani, R (1999). Phylogeny of the Ichthyopterygia. *Journal of Vertebrate Paleontology* **19** (3), 473–496.

Nicholls, E and M Manabe (2004). Giant ichthyosaurs of the Triassic — a new species of *Shonisaurus* from the Pardonet Formation (Norian: Late Triassic) of British Columbia. *Journal of Vertebrate Paleontology* **24** (4), 838–849.

Slaughter, B and B Hoover (1963). Occurrences of ichthyosaurian remains in the Cretaceous of Texas. *Texas Journal of Science* **15.3**, 339–343.

Van Vranken, N (2017). A Note on the Occurrence of an Upper Cretaceous Ichthyosauromorph Tooth from West Texas. *The Journal of Paleontological Sciences* **7** (2).

Zammit, M (2012). Cretaceous Ichthyosaurs: Dwindling Diversity, or the Empire Strikes Back? *Geosciences* **2.2**, 11–24.

# Traces in the terrarium

**Daniel Hembree**

While most people think of fossils as preserved bones, teeth, shells, leaves, or even impressions of entire animals, these body fossils represent only a portion of the life that has existed on Earth. Many organisms would be missing from the fossil record if they had not left behind other evidence of their existence in the rocks themselves: trace fossils.

Trace fossils (or ichnofossils) are the preserved tracks, trails, burrows, borings, bite marks, and faecal material left behind in sediment, soil, rocks, and other media by animals as they engaged in a variety of behaviours (Bromley 1996). These behaviours include locomotion, escape, dwelling, resting, grazing, feeding, nesting, pupation, and more (Seilacher 1986).

I study neoichnology. I document the burrowing behaviours and traces of extant continental organisms. These data are primarily used to better interpret similar trace fossils occurring in continental sedimentary rocks. The identity of the tracemakers and the behaviours involved in the production of trace fossils are typically inferred from the morphology of the trace. These interpretations require knowledge of the types of traces produced by organisms in modern settings.

Laboratory experiments with modern burrowing continental animals allow for the direct observation of the production of traces under controlled environmental conditions. In my laboratory we have investigated several species of millipedes, centipedes, scorpions, whip scorpions, spiders, salamanders, toads, and skinks from a range of environments and ecological niches (Hembree 2016). These animals represent common soil organisms that have a long evolutionary history known primarily from body fossils. While their trace-fossil record is comparatively poor, this is likely due to a lack of recognition rather than a lack of fossil material. The experiments involve observation of the burrowing techniques of each species, documentation of any surface features produced, observation of behaviours associated with traces, and documentation of trace morphologies.

Specimens are placed in sediment-filled terrariums for periods of 7–60 days. The sediment is composed of varying concentrations of medium- to fine-grained quartz or carbonate sand, clay, and organic matter. Specimens from each species are placed in small to large terrariums to test the effects of available space on animal behaviour and burrow morphology. Those animals that exhibit gregarious or communal behaviour are placed in terrariums individually and in groups to evaluate the effects of group behaviour on burrow morphology. The experiments are also designed to expose the animals to different environmental parameters including soil composition, soil density, and soil moisture.



Observational data collected includes the time the enclosure was occupied before burrowing began, the burrowing techniques used, and the time taken to burrow completely into the sediment. When the burrow is completed, daily observations are made to document the behaviours associated with the burrow. At the end of each experimental period, the animals are removed from the enclosure and the open burrows are filled with plaster.

The burrow casts are described both qualitatively and quantitatively. Architectural morphology consists of the burrow's general appearance, dimensions, cross-sectional shape of shafts and tunnels, orientation in the sediment, type and amount of branching, and degree of interconnectedness of different burrow elements. Surficial morphology includes structures (scratches, bumps, linings) on or around the burrow walls produced during excavation and occupation of the burrow.

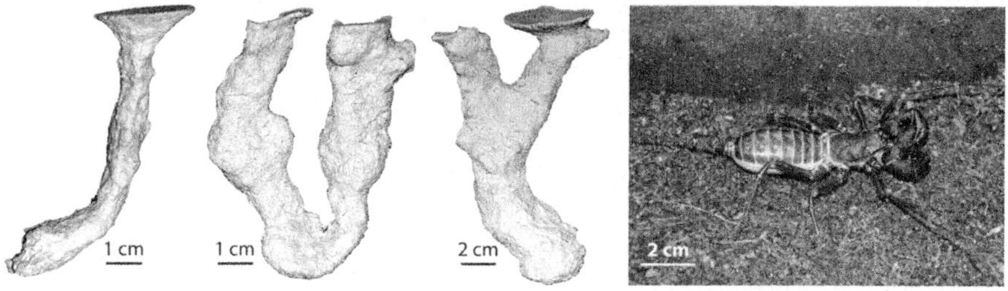


Casts of three different types of burrows (left) produced by the whip scorpion *Mastigoproctus giganteus* (right).

The burrows produced in these experiments include vertical shafts, subvertical ramps, J-shaped burrows, U-shaped burrows, W-shaped burrows, Y-shaped burrows, O-shaped burrows, helical burrows, branching burrows, and mazeworks. In addition, some of the animals produced many different types of sedimentary structures, resulting from their movement through soft sediment. Results of experiments such as these provide analogs with which to compare and interpret trace fossils, making inferences as to potential tracemakers, associated behaviours, and the environments in which they were produced.

## References

Bromley, R (1996). *Trace Fossils: Biology, Taphonomy, and Applications*. Chapman and Hall, London.

Hembree, D (2016). Using experimental neoichnology and quantitative analyses to improve the interpretation of continental trace fossils. *Ichnos* **23**, 262–297.

Seilacher, A (1986). Evolution of behaviour as expressed in marine trace fossils. In M H G Nitecki & J A Kitchell (Eds.), *Evolution of animal behaviour: Palaeontological and field approaches* (pp. 62–87). New York: Oxford University Press.

# Vertebrate palaeontology: more than fossil bones

**John Hutchinson**

While new discoveries of skeletal material are the lifeblood of vertebrate palaeontology, the field has always been about more than digging up and describing and interpreting bones. Today this is more the case than ever. Palaeontology as a whole has diversified and integrated with many other fields. The boundaries are so fuzzy that it can even seem hard to distinguish palaeontology from allied disciplines such as evolutionary biology, comparative anatomy, and geology.

The growing interdisciplinary synthesis has come through the study of a wider array of fossil types, from footprints to skin and other soft tissues, and even from fossilized eggs and excrement to preserved organic molecules from ancient vertebrates. Palaeontologists draw on novel skills — such as medical imaging, dissection, experimentation, and computer modelling — to help interpret those finds alongside skeletal data, and collaborate with other disciplines, especially biology and geology but also engineering, physics, computer science, and other areas.

The application of new techniques to non-skeletal material has led to exciting discoveries and insights once widely regarded as impossible. Most memorably just 20 years ago it was common to say 'we'll never know what colour dinosaurs were,' but now fossilized feather-like filaments have been found for several species of dinosaurs that reveal not only colours but also patterning of colour across the body, so that we can confidently reconstruct their external appearance with a remarkable degree of detail (e.g. Vinther et al. 2008).

In order to interpret the myriad types of evidence they must deal with, vertebrate palaeontologists must be more broadly skilled today than ever. Anatomical dissection of modern animals is commonly used to establish how soft tissues such as muscles and tendons relate to the bones that they attach to (e.g. Witmer 2001). In turn, fossil material is studied to inspect the small, striated 'muscle scars' that reveal where those soft tissues attached to bones, as soft tissues often become mineralized at their interface with bone and thus bones themselves preserve features of soft-tissue anatomy. By interpreting this evidence, which almost every skeleton preserves some of, we can reconstruct the whole-body anatomy of extinct vertebrates in new detail.

Reconstructing how soft-tissue anatomy changed over time — such as how scaly reptiles and other early land animals evolved feathery or hairy skin as they

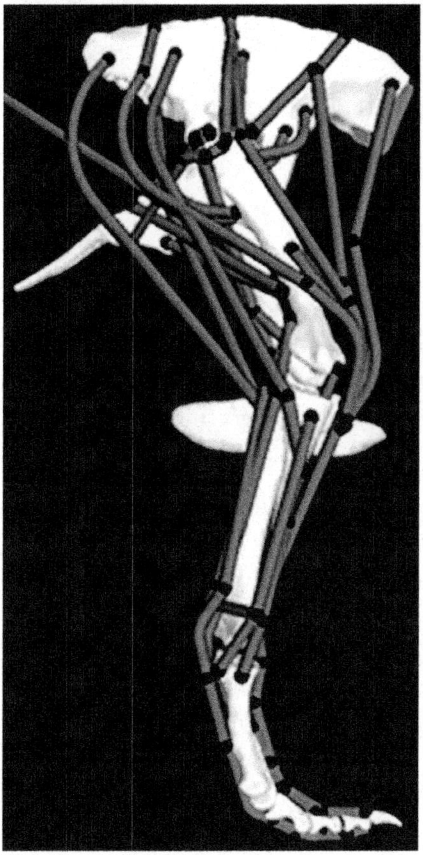

Biomechanical computer model of the hindlimb (right leg) bones and muscles (grey lines) of a *Tyrannosaurus rex*.

evolved into birds or mammals — is interesting in itself. But data on the soft tissues of extinct vertebrates are also valuable for testing tissue function.

Bones alone don't create behaviour — real motions of animals emerge from interactions of the hard and soft tissues of organisms with the nervous system and the environment. With this knowledge in hand, we can create 3D models of extinct vertebrates using medical imaging of fossils, digitally adding muscles and tendons to them using computer graphics software. We estimate how those animals might have stood or moved using biomechanical computer modelling tools adopted from engineering. For example, could *Tyrannosaurus rex* have run over 65 km/h as some palaeontologists had thought 20+ years ago? The integration of evidence from multiple sources such as bones, soft tissues, and modern technologies shows us that *T. rex* probably could not run faster than 25–40 km/h at best (e.g. Gatesy et al. 2009).

The questions now are: how far can vertebrate palaeontology go? What can't we learn about extinct life forms and their evolution? What assumptions do we make now about what is unknowable that in 20 more years we'll find out to be wrong?

## References and acknowledgments

Gatesy, S, M Baeker, and J Hutchinson (2009). Constraint-based exclusion of limb poses for reconstructing theropod dinosaur locomotion. *Journal of Vertebrate Paleontology* **29** 535–544.

Vinther, J, D Briggs, R Prum, and V Saranathan (2008). The colour of fossil feathers. *Biology Letters* **4** (5), 522–525.

Witmer, L (2001). Nostril position in dinosaurs and other vertebrates and its significance for nasal function. *Science* **293** (5531), 850–853.

I thank the Natural Environment Research Council (UK) for funding that supports my research in this area, and collaborators Steve Gatesy, Mariano Garcia, Vivian Allen, and others for their influence.

# What is a fossil? An epistemological view

**Alex Léna**

The purpose of this essay is to explore the space that exists between the moment of a fossil's discovery and its interpretation or insertion into a theoretical framework. To do this, I consider the fossil in terms of three analytical concepts: the object, the fact, and the phenomenon. There is a considerable amount of literature about the use of the fossil but little that considers the constraints that fossils impose on the palaeontologist.

In this essay, I will use the term epistemology to refer to the study of the complex system which allows scientists to create and justify knowledge. My philosophical approach is inspired by the French twentieth-century school of Gaston Bachelard and his heirs Georges Canguillheim, Michel Foucault, and particularly Gilbert Simondon and a new philosophical approach named object-oriented philosophy developed by Quentin Meillassoux and Graham Harman. The concept of an epistemology through the natural object is, as far as I know, original.

## The fossil as natural object

Fossils, as natural objects, are not rare, and have been observed ever since people have been observing the world. We regularly come across the claim that 'fossils are special'. This is not simply a rhetorical trope: fossils are special because of their exceptional status in the epistemological world of natural objects.

There are two kinds of natural object: the visible and the invisible. The latter, such as atoms and molecules, were hypotheses before their existence came to be generally accepted. The former, such as fossils, living animals, or water, were not hypothetical. Nevertheless, we find crucial ontological differences between objects. Some elicit an immediate intuition regarding their properties, for instance the ostensible properties of a living animal. While others, such as fossils keep their nature hidden. The revelation of this hidden nature requires an active intellectual process on behalf of the observer. Even when fossils are in plain view, they are not in 'plain grasp'. Fossils are tangible but their history has been written and re-written — they are palimpsests.

## The definition of a fossil

Reviewing the history of palaeontology with a focus on the concept of the fossil, we see that some important characteristics emerge, as observed by Lamarck (1801) and Pictet (1844). Fossils are organic in origin and have been corrupted by taphonomic processes. The organisms which in turn characterize the fossils are typically extinct. The palaeontologists can not admit recent living organisms within their definition without profoundly modifying the meaning of what is a fossil.

Given their palimpsestic nature, fossil remains are definitively a-phenomenal and inaccessible to experimentation. The palaeontologist needs to recover or reinvent the fact and the phenomenon themselves. While a biologist can directly observe nature, the palaeobiologist needs to proceed via the medium of the fossil. Fossils are the real mediator between the scientist and nature. Prior to becoming an object for biology, fossils are biological archives in a historiographical sense (Léna 2013).

## Form and life as a *princeps* fact

Form and structure are the necessary points of entry for the palaeontologist into his inquiry. But the characterization of a form within the fossil requires an established reference, generally in the present. Recognizing a form in a fossil, whether by analogy or homology, reinforces the living nature of the fossil; form and life are considered as being the same thing.

Form provides the primary grounds for establishing the central, primal fact which is life. Without recognized life there is no fossil, without fossils there is no palaeontology.

Because form is the unique 'point of entry' for the palaeontologist, analogy and homology constitute the first logical inferences in the field. Because the fossil is a-phenomenal, it obliges the scientist to adopt a specific type of logic close to circular reasoning. Indeed, this is the lot of all historians. Because fossils aren't experimentable objects, palaeontologists must develop Sherlock Holmes's core skill: abductive reasoning.

Inductive or deductive inference do not apply in palaeontology, but despite this chain of so-called logical fallacies, palaeontology is always creating new knowledge with a high degree of accuracy. I argue that this philosophical toolbox is the consequence of this historicity. Maybe we should imagine a new field of study — the epistemology of the historical science of nature.

## References


Lamarck, J-B (1801). *Système des animaux sans vertèbres, ou tableau général des classes, des ordres et des genres de ces animaux; présentant leurs caractères essentiels et leur distribution...*, Paris, Detreville, VIII: 1–432

Léna, A (2013). L'archive biologique en question : le paléontologue est-il un historien ou un biologiste ? *Bulletin d'Histoire et d'épistémologie des sciences de la vie* **20** (2), pp. 197–213.

Pictet, F (1844). *Traité élémentaire de paléontologie; ou, Histoire naturelle des animaux fossiles, considérés dans leurs rapports zoologiques et géologiques*, Paris, Langlois et Leclerq.

# What's the point of phylogenies?

**Javier Ortega-Hernández**

I was recently confronted by this question from a geochemist, who was intrigued as to why palaeobiologists like myself spend large amounts of time stressing over the filial relationships between long extinct organisms and their living relatives.

I was rather surprised by the question. After all, most evolutionary biologists would agree that reconstructing the branches that make up the Tree of Life is a fundamentally significant endeavour in itself. Indeed, ever-increasing amounts of data and new methods are continuously being used for elucidating the evolutionary relationships between living beings, as accurately and in as much detail as possible. However, making bigger and better reconstructions of the Tree of Life is not the only — or ultimate — goal of phylogenies, but rather a necessary step in a more holistic process.

Phylogenies are tools that allow us to contextualize the diversity of life, following the premise that all living beings are related to each other by virtue of sharing a common ancestor. However, most organisms possess a mixture of ancestral and derived features that reflect their heritage, function, or a combination of both. By comparing these characters, it is possible to convey how different species relate to each other through an evolutionary tree or cladogram. Therefore, phylogenies serve two main purposes: they represent a classification of biodiversity that is based on ancestry, and they enable us to follow the evolutionary changes that led to the biodiversity itself at any given point in time.

Categorizing things according to their shared attributes is an almost instinctive human behaviour, and thus most people are familiar to some extent with the systematic utility of phylogenies. After all, it intuitively makes sense that animals that possess feathers, or plants that produce flowers, would belong to clearly defined groups that share a single common ancestor (i.e. monophyletic) from which they inherited those particular traits. However, recognizing these so-called monophyletic groups is more often than not problematic. Organisms are highly dynamic entities, and usually posses many characteristics that carry entirely different implications depending on how they are interpreted. For example, most living reptiles have paired openings on the sides of the skull known as temporal fenestra (Latin for *window*), which serve as sites for muscle attachment. Turtles are unique in this regard, as they completely lack tem-

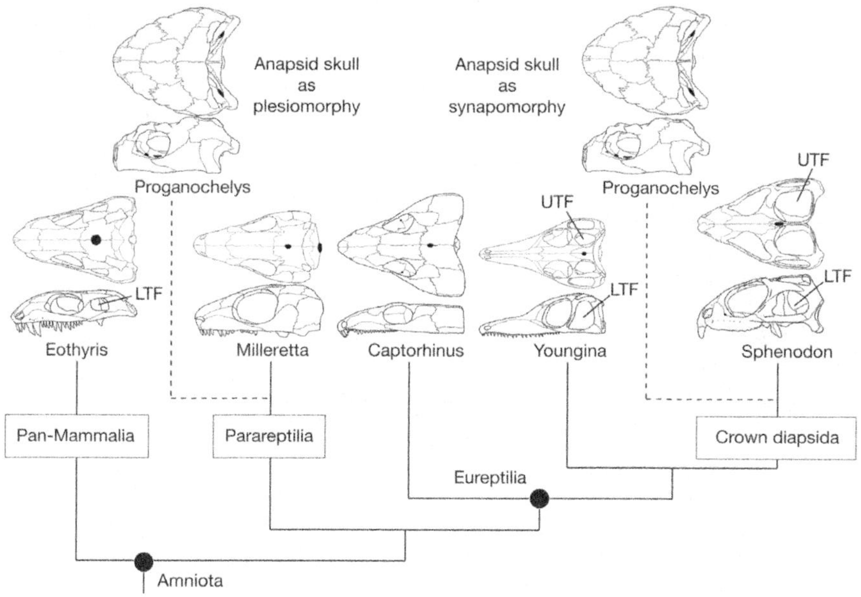


Competing hypotheses for the origin of the anapsid (i.e. without fenestra) skull of turtles, exemplified by the alternative phylogenetic positions of the Triassic species *Proganochelys quenstedti*. Abbreviations: LTF: lateral temporal fenestra; UTF: upper temporal fenestra. (Modified from Bever et al 2015.)



poral fenestra. These observations prompted hypotheses that consider turtles as either ancestral reptiles that primitively lack fenestra — as observed in some fossil taxa — or as highly derived organisms that originally had fenestra but subsequently lost these structures. The two alternatives provide different views of the phylogenetic position of turtles and their close relatives, as well as the transformations that resulted in the evolution of these animals.

The best way to surmount these difficulties is to examine as much data as possible in order to resolve whether the conflicting characters are the result of homology (i.e. shared similarity through common ancestry) or convergence (i.e. similarity as a result of analogous function); hence the considerable efforts invested in obtaining information from as many different species as possible. Returning to our example, the evidence available from phylogenomics — that is phylogenetic trees based on molecular sequence data from a large sample of genes — and recent fossil discoveries indicate that the skulls of early turtles indeed bore fenestra, but that these ancestral openings became secondarily closed later in their history. Critically, it is only possible to obtain insights on the order in which these transformations occur — and by extension the origin of novel body plans — by interpreting the information from individual species in the broader context of their filial relationships. So, the point of phylogenies is to better understand evolution, it's as simple as that.

## References

Bever, G, T Lyson, D Field, and B Bhullar (2015). Evolutionary origin of the turtle skull. *Nature* **525**, 239-242.

# When is a character functional?

**Enrico Savazzi**

In a living organism, the function of a morphological character can be assessed by direct methods. A function hypothesis is formulated, and can be directly proved or falsified by direct observation and measurement. Experimentally altering the character or its adaptive context (often the environment) provides further data. Indirect evidence (see below) can suggest functions, but direct experimental verification is regarded by biologists as the hallmark of good science.

Palaeontologists are forced to use indirect methods to assess the function of morphology in a fossil organism. Does this mean that functional morphology is out of the question in fossils, or can good science be done by indirect methods alone?

Most palaeontologists believe it can. The fossil record makes up for its incompleteness (for example because only skeletal parts are usually preserved) with its higher diversity than the Recent record, and the additional dimension of geologic time. It would not be wise to ignore fossil evidence when studying adaptation and functional morphology.

Among the approaches available to palaeontologists, I suggest the following. Conclusions reached with multiple approaches are inherently stronger, so use as many approaches as feasible in each investigation.

- **Mathematical and physics-based modelling.** This usually means quantitative computer modelling, although pencil-and-paper calculations are possible.
- **Physical models and physical testing.** Test the performance of a model (or the actual fossil if feasible) in the laboratory. Test also modified models as controls. 3D printing has been used in this context.
- **Paradigm method.** Compare the observed morphology with a theoretical model designed to provide the hypothesized function at an optimal level. Do not base the design on the actual organism.
- **Taxonomically related, or homeomorphic, recent counterparts.** If the fossil has recent relatives or convergent forms, study their functional morphology. Remain aware that a similar character can have different adaptiveness in different contexts.

- **Gain supporting information** from palaeoecology, taphonomy, and sedimentology. For example, if taphonomy suggests infaunal life in soft sediment, and you think you see burrowing adaptations, you are probably on the right track. If taphonomy suggests instead sessile life on rock surfaces, you may need to re-think your functional hypothesis.
- **See the character in its context.** A character is never functional by itself. It is only functional in the context of the whole organism.
- **Integrated conceptual frameworks.** Approaches that include the genetic and constructional building-blocks available to the organism, and/or its evolutionary history and relatives, are inherently stronger than studying a single species outside its evolutionary context. Constructional morphology is a foremost example.
- **Allow for the unexpected.** For example, Recent lingulid brachiopods burrow with the pedicle trailing behind the shell. They lack burrowing sculptures, but so do most of their fossil taxa. All is in order when an Ordovician lingulid is found to possess burrowing sculptures that indicate the very same burrowing mechanism. Palaeozoic obolids have no recent relatives, but are similar to lingulids in shape and palaeoecology, and several obolids display a broad variety of burrowing sculptures. What to make of the fact that all obolid sculptures indicate burrowing in the opposite orientation, compared to lingulids? The conclusion: lingulids and obolids burrowed in different ways, and evolved burrowing behaviour independently of each other in spite of their shape similarities (Savazzi 1996).

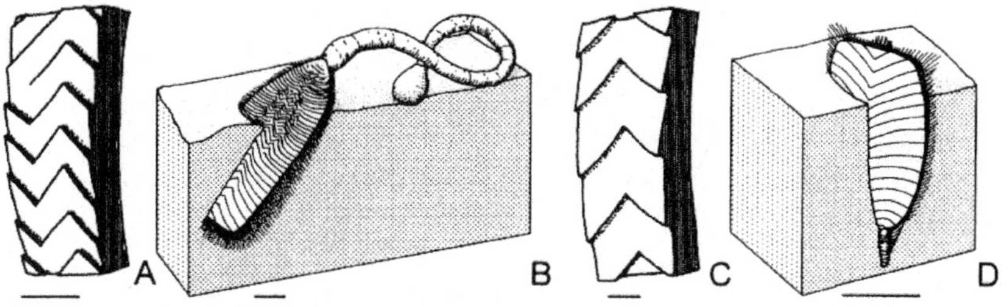


Lingulids (A-B) and obolids (C-D) burrowed in different ways, and evolved burrowing behaviour independently of each other in spite of their shape similarities. The shell is oriented with the pedicle uppermost in A and C.

## References

Savazzi, E. (1986). Burrowing sculptures and life habits in Paleozoic lingulacean brachiopods. *Paleobiology* **12**, 46–63.

# When the present is *not* the key to the past

## Joseph Botting

There's an old adage in geology: the present is the key to the past. This simply means that looking to the modern world helps us unlock the record in the rocks. While this principle is vital to deciphering the fossil record, it's not perfect — especially in the deeper reaches of animal evolution. Sponges, which are often regarded as the most primitive living animals, are a case in point.

You could be forgiven for thinking of sponges as 'primitive'. They have no real organs, no symmetry, and a few of them can even reassemble after being squeezed through cheesecloth. The individual choanocyte cells that create currents through their bodies, and catch food in the process, are remarkably similar to choanoflagellates — our closest living single-celled relatives. If you meet one, it's likely to be a gelatinous blob stuck to rocks on the shore. Indeed, it wasn't until the mid-nineteenth century that they were recognized as animals at all. They were previously thought to be multicellular protists, on the boundary of true animals, but not fully within them. This view is now being overturned, with the evidence coming from both genetics and the fossil record.

Genetically, sponges should be simple. They don't have muscles, or nerves, or even much in the way of bodily organization. In theory they shouldn't have genes for those things, but it turns out that they do. Some sponges have genes that in other animals are used to create body structure, polarity, a nervous system, and muscles. They're usually pretty mangled, but recognizably the same genes. That's extremely interesting, although its exact significance remains to be worked out.

I'm going to focus on the fossil record, though, which is rather mixed. Some modern sponges have solid mineral skeletons (either silica or calcium carbonate) that can easily be fossilized. But most sponges have loose skeletons of minute spicules, bound into a framework by soft tissues, and very rarely preserved intact. So which skeleton came first, if any? The traditional assumption is that the first sponges had no skeleton, and spicules evolved separately in the different classes. But does that stand up to scrutiny?

Tracing back the fossils, the first thing we find is that all the solid, easily preserved skeletons evolved within relatively advanced groups of sponges. Go

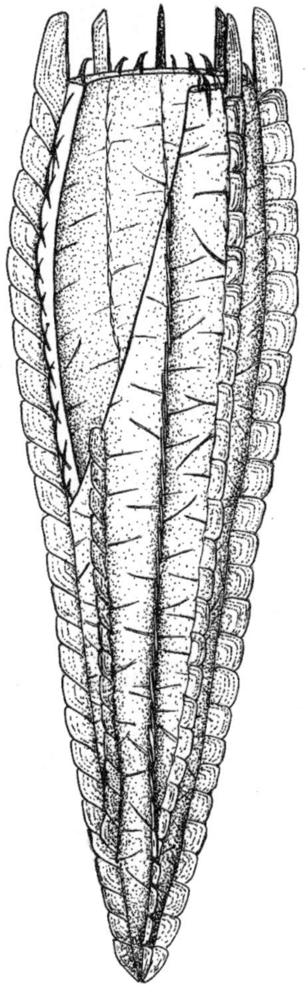

*Takkakawia*, Cambrian

back further down the evolutionary lines, and you have only spicules: tiny needles in demosponges, cross-shaped hexactines in the hexactinellids, and six-spoked 'hexaradiate' calcite spicules in the Calcarea. That's all as it should be, but then things start getting weird. In the Cambrian, many sponges were breaking the rules. We find a combination of hexactines and hexaradiates in the supposed Calcarean *Eiffelia globosa*, spicules made of calcite wrapped around a silica core in the controversial *Lenica*, and hexactines with the internal structure of demosponge spicules in *Cyathophycus*. Suddenly, the divisions between the classes collapse, because we've reached the 'stem groups' of those lineages: the branches before the modern groups diverged.

All of these early sponges are aesthetically far removed from the blobs on modern beaches. These are delicate, vase-shaped animals with exquisite spicules in a precisely ordered array, like the modern glass sponges of the deep sea. Some of these early sponges had four-fold symmetry, something completely unknown among living species. Perhaps most spectacular of all is *Takkakawia* from the Cambrian Burgess Shale of Canada (Botting 2012, left). This is a very long way indeed from a colonial protist.

All of this goes to show: we mustn't judge a sponge by its modernity. Those blobs are highly adapted to life in a challenging environment, and are nothing like their ancestors. It is too easy to see simplicity in modern species, and project backwards to a simple beginning — one that, for sponges, may never have existed.

## References and further reading

Botting, J (2012). Reassessment of the problematic Burgess Shale sponge *Takakkawia lineata Walcott*, 1920. *Canadian Journal of Earth Sciences* **49** (9), 1087–1095.

Dunn, C, S Leys, and S Haddock (2015). The hidden biology of sponges and ctenophores. *Trends in ecology & evolution* **30** (5), 282–291.

# Why do you keep those old fossils?

## Chris Cleal

Those of us working in the museum sector are often asked, why do we keep all those old fossils? OK, a dinosaur skeleton or even a trilobite can look nice on display, but what about those other, less-impressive looking specimens? The scientific paper containing the identifications was published ages ago — why not just go and read the paper?

When anyone asks me why we are keeping some of these old fossils in our basement stores, this is the story I tell them.

A major collection held by National Museum Cardiff is the David Davies Collection of Welsh Carboniferous plant fossils. About 100 years ago, Davies was a colliery manager in South Wales who developed a passion for palaeobotany (Thomas 1986). Initially his collecting rather lacked focus but this changed when he met the flamboyant British palaeobotanist Marie Stopes. In 1911, Stopes had spent some time in Canada where she had started to investigate Carboniferous plant ecology (Stopes 1914). On returning to Britain she heard about Davies and went to visit him at his home in Gilfach Goch. She apparently causing a stir by walking around his village in trousers, but more importantly explained to him how his systematic collecting could be used for palaeoecology. Stopes did little subsequent geological work — she shifted her attention to family planning — but her influence on Davies was profound and resulted in him embarking on a campaign to collect systematically from all of the main South Wales coals. The result was a vast collection of specimens, on which he based his pioneering paper on coal palaeoecology published by the Royal Society (Davies 1929).

Davies not only collected extensively but also recorded in his notebooks every specimen that he saw. The result is data based on a series of assemblages of up to 30 000 specimens each, reflecting vegetation change in South Wales over some 10 million years. The Late Carboniferous world has many similarities to today and we can use data from this time to test some of our ideas about how Earth's processes work, such as those relating to climate change.

An analysis published in 2007, based mainly on Davies's data, revealed that diversity in the South Wales coal swamp vegetation remained remarkably stable for the first 5 Ma (until Middle Moscovian times). Plant diversity then declined significantly, mainly due to a loss of spore-plant species (club-mosses, horsetails, and ferns) and then a little later (Late Moscovian times) it rebounded through the appearance of marattialean ferns and medullosalean pteridosperms. But what caused these changes — climate or habitat-drainage?

The idea that Carboniferous vegetation and climate were linked has been around for years, but the question has always been about causality. Did climate cause vegetation change, or did vegetation change alter the climate? By comparing Davies's data with those from elsewhere in Europe and North America (through collaborative projects within the International Geosciences Programme — IGCP 469 and 575) we have been able to show that this question is too simplistic. Vegetation and climate were linked through a complex set of feedback mechanisms — again perhaps not a totally surprising idea, but now supported by empirical data. Importantly, the change seems to have started in eastern Europe, in Early Moscovian times, where landscape changes triggered by tectonics caused a change to a more fern-dominated vegetation. This reduced overall rates of transpiration of the swamp vegetation (the throughput of water in the ferns was less than in the giant club-mosses that previously dominated the swamps), which in turn reduced rainfall. The reduced rainfall caused the fern-dominated vegetation to spread into adjacent areas, which then caused further reductions in rainfall. The resulting feedback between climate and vegetation change progressively rolled across the rest of Europe, finally reaching North America some 3–4 Ma later, at the end of the Moscovian age (Cleal et al. 2011).

So, the data accumulated about a century ago by a colliery manager in South Wales, who had been inspired by a remarkable lady palaeobotanist, is throwing new light on how Late Carboniferous vegetation, atmosphere, and climate interacted. And importantly, this is backed up by empirical data in a museum collection.

## References and further reading

Cleal, C (2007). The Westphalian–Stephanian macrofloral record from the South Wales Coalfield. *Geological Magazine.* **144**. 465–486.

Cleal, C, S Opluštil, B Thomas, and Y Tenchov (2011). Pennsylvanian vegetation and climate in tropical Variscan Euramerica. *Episodes* **34**. 3–12.

Davies, D (1929). Correlation and palæontology of the Coal Measures in east Glamorganshire. *Philosophical Transactions of the Royal Society of London, Series B* **217**. 91–153.

King, S, C Cleal, and J Hilton (2011). Common ground between two British Pennsylvanian wetland floras: using large, first-hand datasets to assess utility of historical museum collections. *Palaeogeography, Palaeoclimatology, Palaeoecology* **308**. 405–417.

Stopes, M (1914). The "Fern Ledges" Carboniferous flora of St. John, New Brunswick. *Canada Geological Survey Memoir* **41**. 1–142.

Thomas, B (1986). *In search of fossil plants: the life and works of David Davies (Gilfach Goch)*. National Museum of Wales, Cardiff, 54 p.

# List of contributors

**Markus Aretz** studied geology and palaeontology at the Universität zu Köln, Germany and the Université de Liège, Belgium. He completed both his masters and his PhD in Köln. Before joining the Université Paul Sabatier, Toulouse, as associate professor for palaeontology, he worked at the Queensland University of Technology and Universität zu Köln. His research is focused on different palaeontological, geological, and stratigraphical topics, often in Carboniferous and Devonian strata. From 2008–16, he served as secretary of the Subcommission on Carboniferous Stratigraphy (SCCS/ICS), and he currently heads the task group for the revision of the base of the Carboniferous.

*Corals meet reefs in the Carboniferous* 34

**Benni Bomfleur** is a palaeobotanist at the Westfälische Wilhelms-Universität Münster (Germany). The focus of his research is plant life of the Mesozoic, but his interest for the unusual has led to some bizarre fossil discoveries, including Jurassic-age chromosomes and the 50-million-year-old fossil worm sperm that spawned the idea for the present essay.

*A closer look at fossil sex* 16

**Joseph P Botting** has worked in Cambridge, London, Leeds, and China. He currently holds honorary positions at the National Museum of Wales, Cardiff, and the Nanjing Institute of Geology and Palaeontology, China, while living mostly in the middle of Wales, working as a freelance educator, entomologist, and harpist. Around earning a living, he continues his research into sponges and exceptionally well-preserved ancient ecosystems, while encouraging local amateur palaeontology. He's married to another palaeontologist, with whom he co-authors lots of papers.

*When the present is* not *the key to the past* 114

**Euan Clarkson** studied geology at the University of Cambridge and has had a long career as a palaeontologist at the University of Edinburgh, Scotland. He has authored 100+ papers and other publications including the book *Invertebrate Palaeontology & Evolution*. Clarkson was president of the Edinburgh Geological Society (1985–87), a trustee of the Natural History Museum (1987–92), and president of The Palaeontological Association (1998–2000). He was awarded the Geological Society of London's Coke Medal in 2010 and the Lapworth Medal of the Palaeontological Association in 2013.

*Scottish Carboniferous Fossil-Lagerstätten* 94

**Chris Cleal** has been working on Carboniferous palaeobotany for over 40 years, variously at the University of Sheffield, the Saarbrücken Mining School (Germany), the Nature Conservancy Council and now at National Museum Wales, where he is now Head of Botany.

**Alex Cullum** is a geologist, palaeontologist, and author currently working in exploration for Statoil in Stavanger, Norway. He studied at Aberystwyth University in Wales and his PhD was on the biostratigraphic correlation of sequences in Arctic Canada. Psychology and how we communicate science and the things that matter are the subject of his latest novel.

**Tony Doré** is a Senior Advisor at Statoil with a global exploration remit. He was previously VP of Exploration North America. Tony has published more than 50 peer-reviewed papers, and edited six books. He has served academia and industry in various capacities, and in 2010 was appointed as an Officer of the Most Excellent Order of the British Empire (OBE) for services to geology. He shares a house in London with his wife and far too many guitars.

**Robin Dyer** is a consultant biostratigrapher specializing in North Sea stratigraphy. Following a BSc in geology, he gained MSc and PhD degrees in micropalaeontology from Aberystwyth and Keele before working for Robertson Research and Britoil. He co-founded GeoStrat in 1986 and has since worked for a large number of operators throughout the Northern North Sea and North Atlantic margin.

**Denver Fowler** is originally from Stockport, UK. He received his PhD under Dr. Jack Horner at the Museum of the Rockies, Bozeman, Montana and is currently employed as curator of palaeontology at Dickinson Museum Center, North Dakota. His research mixes stratigraphy with dinosaur evolution and behaviour, and he spends many months each year conducting fieldwork in the Western United States, collecting new dinosaur fossils.

**David A T Harper** is a geologist and palaeontologist, specializing in fossil brachiopods and numerical methods in palaeobiology. He is professor of palaeontology in Earth Sciences and Principal of Van Mildert College, Durham University. He is a foreign member of the Royal Danish Academy of Sciences and Letters, the Royal Swedish Physiographic Society, and is an Einstein Professor in the Chinese Academy of Sciences. He is currently Chair of the International Commission on Stratigraphy.

**Daniel I Hembree** is an associate professor in the Department of Geological Sciences at Ohio University in Athens, Ohio. He earned his MS and PhD degrees from the University of Kansas. His research interests primarily lie with animal–substrate interactions in ancient and modern continental environments. Past and current research projects involve the study of the influence of climate changes on ancient soils and soil ecosystems. His research also involves the study of burrowing behaviours and resulting biogenic structures of extant terrestrial annelids, arthropods, amphibians, and reptiles in the laboratory and field.

*Traces in the terrarium* 104

**Liam Herringshaw** is a palaeoecologist interested in animal–sediment interactions through time and space, particularly in marine environments. A lecturer in geology at the University of Hull, Liam has a BSc in geology and physical geography from the University of Liverpool and a PhD in palaeobiology from the University of Birmingham. Before moving to Hull in 2015, Liam worked as a researcher at the University of Aberdeen, Memorial University of Newfoundland, and Durham University, as well as a freelance writer, editor, and tutor.

*Earth's own asteroids* 44

**Wyn Hughes** has MSc, PhD, and DSc degrees from Aberystwyth University and is a micropalaeontologist and microfaciesologist, and director of Applied Microfacies Ltd. He has experience from the Solomon Islands Geological Survey, Robertson Research International, and Saudi Aramco. He is an associate of the Natural History Museum, visiting lecturer at the universities of Birmingham and Urbino, and adjunct professor at King Fahd University, Saudi Arabia.

*Micro-dung and its uses* 78

**John R Hutchinson** is a professor of evolutionary biomechanics whose research focuses on how animals work, especially locomotion, and how those mechanisms have evolved. His team's studies have reconstructed how tyrannosaurs and other dinosaurs may have moved, shown how elephants use a bizarre bouncing run, and discovered that the earliest vertebrates to move on land might have only been able to manage a slow, forelimb-driven 'crutching' gait similar to modern mudskippers.

*Vertebrate palaeontology: more than fossil bones* 106

**Frans Jorissen** studied geology at Utrecht University and earned a PhD in 1988 on the ecology of modern benthic foraminifera in the Adriatic Sea. In 2001, after 10 years at Bordeaux University, he obtained a chair at Angers University and became director of the Laboratory of Recent and Fossil Bio-Indicators, probably the largest centre of foraminiferal research in the world. His research has focused on foraminiferal ecology, palaeoceanographical proxy development, and more recently, on the application of foraminifera in environmental quality studies.

*Foraminiferal microhabitats* 46

**Benjamin P Kear** is curator of vertebrate palaeontology at the Museum of Evolution in Uppsala University, Sweden. His first introduction to palaeontology was by studying fossil kangaroos at the University of New South Wales in Sydney, Australia. Since then his PhD and postdoctoral research have taken him through the disparate fields of Mesozoic polar marine ecosystems and the molecular systematics of modern Australasian marsupials. Today, he still pursues these fascinating topics (albeit with entertaining excursions into fossil turtles, dinosaurs, bony fish, etc), and maintains major field exploration programs in Australia, Saudi Arabia, and Greenland.

*How kangaroos got their bounce* 58

**Hans Kerp** is a palaeontologist with a passion for fossil plants. He received his PhD in geology from Utrecht University in 1986. In 1991 he was appointed professor of geology and palaeobotany at the University of Münster. His research on the Early Devonian Rhynie chert and on Late Palaeozoic–early Mesozoic terrestrial ecosystems has so far resulted in more than 120 peer-reviewed papers. Since 1994 he has been the editor of one of the leading journals in palaeobotany and palynology.

*A closer look at fossil sex* 16

**Steffen Kiel** is a palaeontologist interested in deep-sea ecosystems, in molluscs, and in what has driven evolution on geological timescales. After completing his PhD at the University of Hamburg, he spent his post-doc years at the Smithsonian Institution in Washington, DC, and at various universities in England, Austria, and Germany. He is now a senior scientist at the Swedish Museum of Natural History in Stockholm. When he is not exploring the fossil record, he is kayaking in the archipelago just outside Stockholm, or can be found in a café reading Thomas Pynchon.

*A taste for bones* 28

**Alex Léna** began his career as a field and lab zoo-archaeologist. A few years later, he integrated geology at the University of Lyon. His encounter with epistemology is not recent, but has crossed all his palaeontological training and scientific practice. Doing science was for him a pretext to studying the philosophy of science.

*What is a fossil? An epistemological view* 108

**David Loydell** is a lecturer at the University of Portsmouth. When not teaching he undertakes research into almost all things Silurian, but with a focus often on graptolites, hydrocarbons, and environments. He received his BSc and PhD from the University of Wales, Aberystwyth. He is a voting member of the International Subcommission on Silurian Stratigraphy.

*Integrated Silurian biostratigraphy* 66

**Gunn Mangerud** is a professor in biostratigraphy and head of the Department of Earth Sciences at the University of Bergen, Norway. She received her masters from NTNU, Trondheim, and her PhD from the University of Bergen. She started her career as a biostratigrapher at the Continental Shelf Institute (SINTEF) in 1988 and moved to Norsk Hydro where she worked in different positions and later as chief geologist. After moving back to academia in 2009, her main interest has been biostratigraphy, with a particular focus on the Carboniferous and Triassic of the Arctic areas.

*Remove the rock to reveal the past* 92

**Dave Martill** is a reader in palaeobiology at the University of Portsmouth, UK, where he specializes in Cretaceous vertebrate faunas and palaeoenvironments. His main interest is in the evolution and diversification of pterosaurs and other reptiles, especially from Gondwana.

*A snake with four legs* 24

**Roger Mason** took up a career as a geologist after discovering *Charnia masoni*. He graduated with a BA from Cambridge in 1963 and continued to a PhD in 1967. He taught metamorphic geology at UCL until 1989, then part-time at Birkbeck, University of London, and University of Bedfordshire. Since 1996 he has taught metamorphic geology at China University of Geosciences, Wuhan City, China.

*In defence of the shoehorn* 62

**Stephen McLoughlin** obtained his PhD from the University of Queensland, Brisbane. He has subsequently held research and teaching positions at the University of Western Australia, The University of Melbourne, and the Queensland University of Technology. Since 2007 he has been a senior curator of palaeobotany at the Swedish Museum of Natural History in Stockholm. His research focuses primarily on Permian to Cretaceous floras of the Southern Hemisphere and northern Europe.

*Antarctica's* Glossopteris *forests* 22

**Peter H Morris** started his career in the oil industry with BP and developed research interests in Triassic–Cretaceous micropalaeontology of northwest Europe and the Middle East. Now, as an independent consultant, he continues to work in these areas as well as applied aspects of megaspore biostratigraphy. He is currently researching the Middle Jurassic of North Yorkshire and Triassic of Hopen.

*Deep in the mire at Yons Nab* 38

**Alison Murray** has a PhD in vertebrate palaeontology from McGill University, Canada. She spent over a decade at the Canadian Museum of Nature in Ottawa, working on both living and fossil fishes before moving west. She is currently a professor at the University of Alberta. Most of her current research is on Cenozoic tropical freshwater fish fossils from Africa and Asia, and Cretaceous marine fishes from Canada.

*Fossil fishing needs patience* 48

**Jon Noad** is a sedimentologist and exploration geologist. After working in gold mining and undersea surveying, he completed a PhD in Borneo and then joined Shell. He spent 18 years in the oil industry, and is now a geology trainer and consultant.

**Michael Oates** gained a geology degree and PhD from University College, London, and spent his career in the petroleum business, the last 13 years as an expatriate in Kazakhstan, India, and Norway. Recently retired from a quarter century with BG Group (just in time), Michael now lives on the Mesozoic of North Lincolnshire, curating a lifetime's worth of fossils in his own private museum, learning about the local geology, and still playing a part on the Geologists' Association council and Rockwatch committee.

**Javier Ortega-Hernández** received his first degree in biology from the Universidad Nacional Autónoma de México, and then moved to the UK to complete an MSc in palaeobiology at the University of Bristol. In 2009 he moved to Cambridge to pursue a PhD in Earth Sciences under the supervision of Dr. Nicholas Butterfield, and in October 2013 became a research fellow at Emmanuel College and the Department of Earth Sciences. Javier became a Herchel Smith Fellow in Biological Sciences at the Department of Zoology at Cambridge in 2016.

**Niall Paterson** is a researcher at the University of Bergen, Norway. He gained his BSc from The University of Glasgow and PhD from Trinity College Dublin. In 2009, Niall joined ExxonMobil Exploration as a biostratigrapher. Niall returned to academia in 2012 and currently specializes in the Middle to Late Triassic palynology of the Norwegian Arctic.

**Tim Palmer** was a Smithsonian Junior Research Fellow, a demonstrator in geology at Oxford, and a senior lecturer at the University College of Wales Aberystwyth. Until early 2016 he was the executive officer of The Palaeontological Association.

**David Peris** studied biology at the Universitat de València, Spain. After an MSc in evolutionary biology he completed his PhD from the Universitat de Barcelona, Spain, in 2015. He worked as a high-school teacher for a while before getting a position as a research assistant at the Universitat Jaume I, Spain, where he is trying to combine his new job in a project related with Acari pests in crops with the study of fossil beetles in amber.

**Vincent Perrier** is a lecturer in geology and palaeontology at the University of Lyon (France). In his research, he focuses on fossil arthropods. After a PhD on the origin of planktonic ostracods in Lyon, he worked at the University of Tartu (Estonia) on the impact of environmental crises on ostracods, and at Leicester University (UK) on the history of zooplankton.

**George Poinar Jr** received his PhD from Cornell University and then took a position at the University of California, Riverside, before transferring to the University of California, Berkeley. Upon early retirement, he became associated with the entomology and integrated biology departments at Oregon State University.

**Mike Pole** researches the history of life on New Zealand's islands. He enjoys travelling, having been to nearly 80 countries. He has taught Australian ecology to American students at the University of Queensland and his work as an exploration geologist has taken him to Mongolia, Indonesia, Turkey, and Tanzania. .

**Rachel Preece** studied geology at University College Wales, Aberystwyth, followed by a PhD at University College London. An industrial internship in California turned Rachel's attentions to petroleum geology, and she joined Chevron in London. Rachel has spent most of her professional career overseas developing her understanding of the petroleum industry and learning through a series of cross-functional and leadership roles. She currently lives in Aberdeen with her young family where she manages the Greater Britannia Area subsurface assets. In her spare time she encourages her children to play outside and get dirty hoping they too will develop a healthy curiosity for rocks and fossils.

**Jérôme Prieto** completed his PhD at the Ludwig-Maximilian University of Munich, Germany, with a focus on the palaeoecology of the small mammals in the Miocene of southern Germany. For the last 10 years he has extended his interest to integrated stratigraphic research in the European Miocene and the consequences for climate and faunal evolution. He also contributes to the study of the poorly documented Tertiary continental palaeoecosystems from Vietnam and Laos.

**Imran A Rahman** is a research fellow at the Oxford University Museum of Natural History, UK. His research is focused on the palaeobiology and evolution of echinoderms. He studies early fossil echinoderms with the aid of X-ray computed tomography and computer simulations, shedding light on their anatomy, functional morphology, and evolutionary relationships.

**Enrico Savazzi** holds a PhD from Uppsala University and has been a palaeontologist at German, Italian, Japanese, Swedish, and US institutions for about two decades. For several years he has been a technical writer in the IT and telecommunications industry, but continues to publish in palaeontology as a hobby, especially on extreme adaptations, puzzling evolutionary histories, and other subjects that catch his curiosity.

**John Scannella** received his PhD in Earth Sciences from Montana State University. His research focuses on the evolution and growth of dinosaurs. He is currently the Interim Curator of Paleontology at the Museum of the Rockies in Bozeman, Montana.

**Iain Scotchman** has a BSc in geology from Leicester University and a PhD in sedimentary geochemistry from Sheffield University. He trained as a geochemist from 1989–90 at Amoco's Tulsa Research Center. He has worked with consultants Gaffney, Cline & Associates, Amoco (UK), Enterprise Oil, and Statoil (UK). He has worked in the North Sea, West of Britain, Ireland, Faroes, and offshore Brazil, as well as unconventional hydrocarbons in North and South Africa, Middle East, Europe and China. He returned to conventional oil exploration in 2013 working the Faroes, Ireland Porcupine Basin Round and North Atlantic Margins evaluation. He retired in March 2017 and now works as a consultant geochemist.

**Robert W Scott** is president of Precision Stratigraphy Associates providing quantitative solutions to stratigraphic problems and training in sequence stratigraphy. He is also a research associate at the department of geosciences at the University of Tulsa, directing student research on carbonate reservoirs and palaeontology. He has taught geology for 25 years. He was a research geologist at Amoco Production Company for 20 years where he provided stratigraphic and palaeontologic data on Mesozoic and Cenozoic rocks. He earned his PhD at the University of Kansas and master and bachelor degrees at the University of Wyoming.

**Mike Simmons** is Technology Fellow for Geosciences at Halliburton with a responsibility to investigate innovation in geoscience as applied to the hydrocarbon exploration process. Previously he was Earth Model Director at Neftex and before that worked at BP, Aberdeen University, and CASP. His main interests are applied stratigraphy and the geology of the Tethyan region. He is a Research Associate at the Natural History Museum and teaches at a number of universities.

**Andrew Taylor** is a consultant reservoir geologist with a specialist interest in ichnology, sedimentology, and biostratigraphy. He has a masters in micropalaeontology from UCW Aberystwyth and a PhD in ichnology and sedimentology from Manchester University. Thereafter he co-founded Ichron Ltd which was sold after 21 years in business. He recently started a new consultancy, Skolithos Ltd, that is focused on providing technical expertise to the hydrocarbon industry and field-based training courses in such places as Portugal, India, and Yorkshire.

**Jon Tennant** completed his PhD at Imperial College London, where his research looked at deep-time evolutionary patterns in groups like dinosaurs and crocodiles. He currently works as a *PLOS Paleo* community editor and is Communications Director for ScienceOpen, a freelance science writer, and author of the kids' dinosaur book *Excavate Dinosaurs!* He can be found on Twitter at *@Protohedgehog*, where he talks about palaeontology and the importance of public access to science.

**Barry Thomas** is a former Head of Biological Sciences and Dean of Science at Goldsmith's College, London, the Keeper of Botany at the National Museum of Wales, Cardiff, and is now an Honorary Professor at Aberystwyth University. He has a DSc from Reading University where he first started his research on fossil lycophytes over 50 years ago.

**Nathan Van Vranken** holds a masters in geology and BS in interdisciplinary studies with an emphasis in palaeontology. Currently, he is based out of the Dallas–Fort Worth area in Texas where he excavates and prepares fossils as a freelance palaeontologist, and does public outreach though social media and at professional conferences. He was one of the scientific advisors on the upcoming video game *Saurian*. His research interests deal mostly with better understanding the evolutionary relationships of marine reptiles, such as mosasaurs, and the Cretaceous marine biostratigraphy of Texas.

**Mike Whitaker** has a degree in geology and biology and an MSc and PhD in palynology. He has worked as a palynologist, stratigrapher, and geologist based in the UK, Holland, Norway, and France, with National Coal Board UK, Shell (UK, International, and Norske), Saga, Norsk Hydro, APT, and Detnorske Petroleum (France). He has assembled his views on the physical and biological evolution of the world from his observations as a palynologist and geologist and through intense discussion.

**Alun Williams** is a petroleum geologist with a PhD in carbonate sedimentology from the University of Aberdeen. He joined the oil industry in 1999, and has worked for Statoil since 2007, where he is currently Exploration Manager for UK licences, and prior to that Leading Advisor for carbonate geology. He still enjoys looking at rocks in his spare time.

**Gert Wörheide** received his PhD (Dr. rer. nat.) in geobiology from the University of Göttingen in Germany in 1998. He has been working on the evolution, phylogeny, and biodiversity of marine sponges and biomineralization of marine invertebrates for more than 20 years. Since 2008 he has held the chair for Palaeontology and Geobiology at the Ludwig-Maximilan University of Munich, Germany, where he applies molecular palaeobiology approaches, including genomics, transcriptomics, and phylogenomics, to unravel animal evolution in deep time.

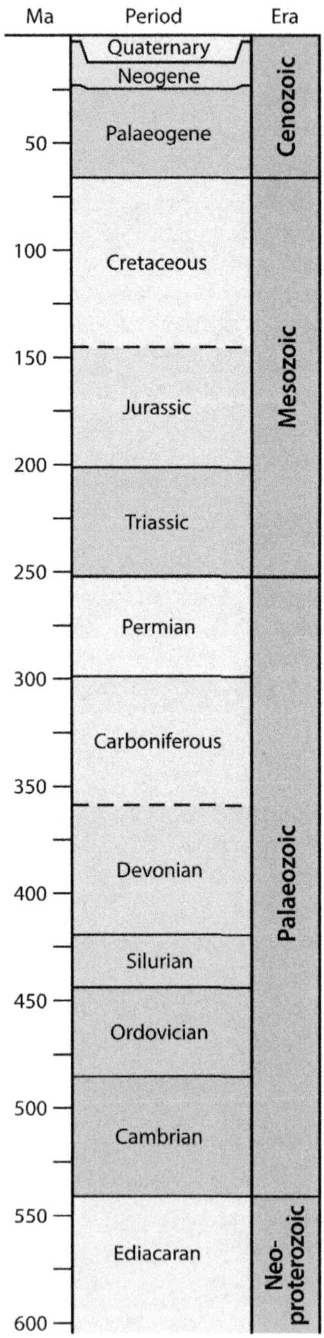
Ma
Period
Era
Quaternary
Neogene
Palaeogene
Cretaceous
Jurassic
Triassic
Permian
Carboniferous
Devonian
Silurian
Ordovician
Cambrian
Ediacaran
Cenozoic
Mesozoic
Palaeozoic
Neo-proterozoic
50
100
150
200
250
300
350
400
450
500
550
600

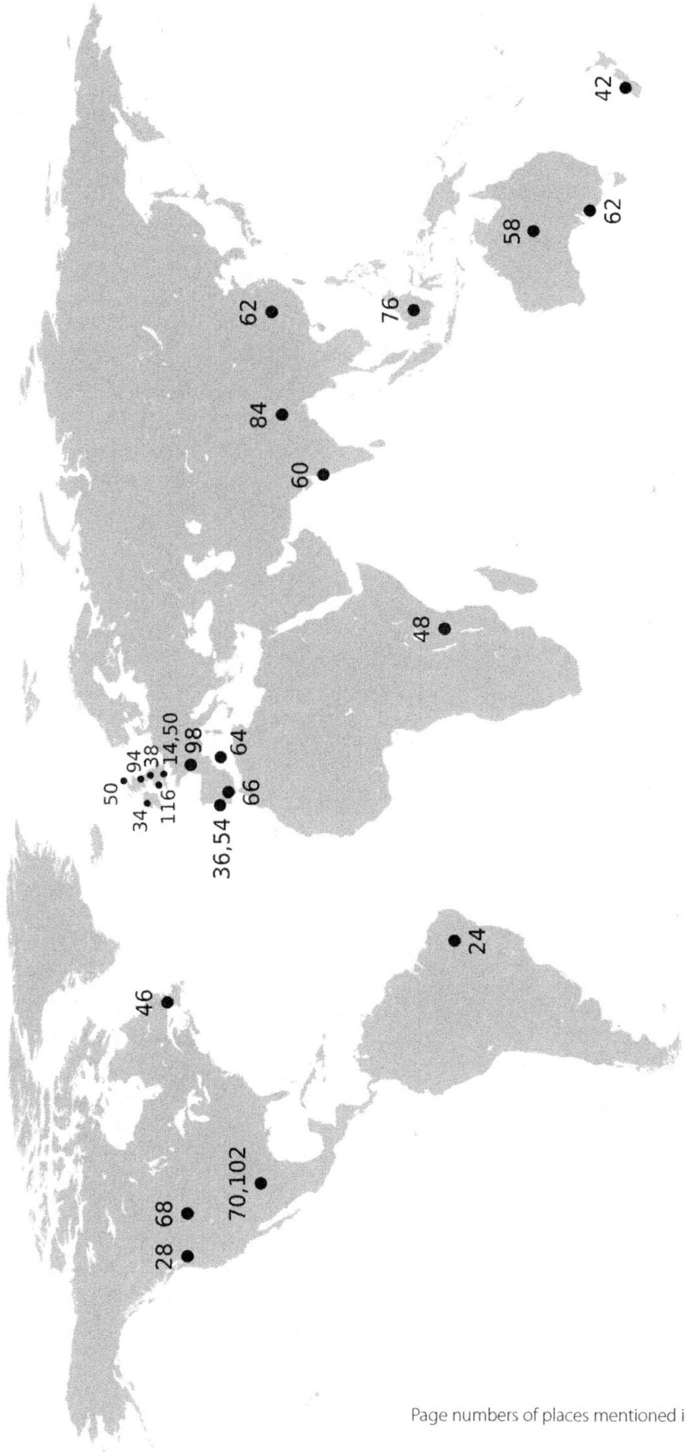


Page numbers of places mentioned in the text.

# Index

**Y**